MEMOIRS
of the
American Mathematical Society

Number 460

Quotients of Coxeter Complexes and *P*-Partitions

Victor Reiner

January 1992 • Volume 95 • Number 460 (second of 4 numbers) • ISSN 0065-9266

American Mathematical Society
Providence, Rhode Island

1991 *Mathematics Subject Classification.*
Primary 51F15; Secondary 05A99.

Library of Congress Cataloging-in-Publication Data

Reiner, Victor, 1965–
    Quotients of Coxeter complexes and P-partitions/Victor Reiner.
    p. cm. – (Memoirs of the American Mathematical Society, ISSN 0065-9266; no. 460)
    Includes bibliographical references.
    ISBN 0-8218-2525-9
    1. Coxeter complexes. 2. Partitions (Mathematics) I. Title. II. Series.
QA3.A57 no. 460
[QA608]
510 s–dc20   91-36297
[516.3'5]   CIP

Subscriptions and orders for publications of the American Mathematical Society should be addressed to American Mathematical Society, Box 1571, Annex Station, Providence, RI 02901-1571. *All orders must be accompanied by payment.* Other correspondence should be addressed to Box 6248, Providence, RI 02940-6248.

SUBSCRIPTION INFORMATION. The 1992 subscription begins with Number 459 and consists of six mailings, each containing one or more numbers. Subscription prices for 1992 are $292 list, $234 institutional member. A late charge of 10% of the subscription price will be imposed on orders received from nonmembers after January 1 of the subscription year. Subscribers outside the United States and India must pay a postage surcharge of $25; subscribers in India must pay a postage surcharge of $43. Expedited delivery to destinations in North America $30; elsewhere $82. Each number may be ordered separately; *please specify number* when ordering an individual number. For prices and titles of recently released numbers, see the New Publications sections of the NOTICES of the American Mathematical Society.

BACK NUMBER INFORMATION. For back issues see the AMS Catalogue of Publications.

MEMOIRS of the American Mathematical Society (ISSN 0065-9266) is published bimonthly (each volume consisting usually of more than one number) by the American Mathematical Society at 201 Charles Street, Providence, Rhode Island 02904-2213. Second Class postage paid at Providence, Rhode Island 02940-6248. Postmaster: Send address changes to Memoirs of the American Mathematical Society, American Mathematical Society, Box 6248, Providence, RI 02940-6248.

10 9 8 7 6 5 4 3 2 1   97 96 95 94 93 92

Contents

Contents

ABSTRACT

We study quotients of the sphere by a subgroup G of a finite reflection group W, investigating both the combinatorial and topological structure of these quotients as certain kinds of cell complexes (*balanced simplicial posets*). In particular, we give sufficient conditions on G for the quotient to be *Cohen-Macaulay* or *Gorenstein* over a field k, and a simple characterizations of those G for which the quotient is a *pseudomanifold*, and when it is *orientable* as a pseudomanifold.

We then look at quotients for particular classes of subgroups G, namely *reflection subgroups*, *alternating subgroups* of reflection subgroups, and their *diagonal embeddings* in the product groups W^r. Here our methods require an extension of some of the theory of *P-partitions* [St3], and *multipartite P-*partitions [GG,Ge2] from the symmetric group S_n to other finite reflection groups, and we present two independent applications of this theory. We then use this extended P-partition theory to show that for the above mentioned groups, the quotient is always *partitionable*, that in some cases it is *shellable*, and when shellable it is either a *sphere* or a *disk*. For all of these groups, the partitioning yields combinatorial interpretations for certain non-negative integers associated to the quotient known as the *type-selected Möbius invariants*. Applications to calculating invariant polynomials of permutation groups and their *Hilbert series* (as developed by Garsia and Stanton [GS]) are also discussed.

Keywords: Coxeter group, Coxeter complex, P-partitions

Chapter 1

Introduction

Finite Coxeter complexes form a class of highly symmetrical triangulations of spheres, and hence provide us with a wealth of interesting and well-behaved finite group actions on spheres. Our study of quotients of Coxeter complexes is motivated by the following problem (considered by Garsia and Stanton in [GS]): Given a subgroup G of the symmetric group S_n acting on the ring $\mathcal{R} = \mathbf{Q}[x_1, \ldots, x_n]$ of polynomials in n variables with rational coefficients, can we explicitly describe $\mathcal{R}^G$, the subring of polynomials invariant under G? Their method proceeds roughly as follows:

1. Replace $\mathcal{R}$ by a related ring $\mathcal{S}$ and show that an explicit description of $\mathcal{S}^G$ leads to one for $\mathcal{R}^G$. The ring $\mathcal{S}$ turns out to be the Stanley-Reisner ring of the *Coxeter complex* of the symmetric group S_n (with

[1]Received by the editor June 1, 1990. This material appeared as part of the author's doctoral thesis at M.I.T. under the supervision of R. Stanley

1

its usual set of Coxeter generators).

2. Get an explicit description of $\mathcal{S}^G$ by decomposing (in a certain fashion) the *quotient of the Coxeter complex* of S_n under the action of G.

They also showed that for Coxeter groups W other than S_n of a certain type (Weyl groups), there exists a ring $\mathcal{R}_W$ analogous to $\mathcal{R}_{S_n} = \mathbf{Q}[x_1, \ldots, x_n]$ in the following sense: Given G a subgroup of W, if one can decompose the quotient of the Coxeter complex under the action of G, one gets an explicit description of the subring $\mathcal{R}_W^G$ of G-invariants. They then proceeded to find such a decomposition (and hence solve the original problem) for subgroups G of W of a certain type (*standard parabolic subgroups*).

Our aim has been to study these quotients of Coxeter complexes in themselves, with the hope of eventually enlarging the class of subgroups G admitting such a solution.

In Section 2, we begin by introducing the main characters of our story, the *Coxeter complex* $\Sigma(W, S)$ and its *quotient* $\Sigma(W, S)/G$ by any subgroup G of W. We explain how $\Sigma(W, S)$ carries the natural structure of a *balanced simplicial complex*, and consequently that $\Sigma(W, S)/G$ is naturally a balanced *simplicial poset* ([St2]). In order to state general results about $\Sigma(W, S)/G$, we define the notions of when a simplicial poset is *Cohen-Macaulay* over a field k, *Gorenstein* over a field k, a *pseudomanifold (with and without boundary)*, or an *orientable* pseudomanifold. The main results of Chapter 1 may then be summarized by saying that $\Sigma(W, S)/G$ is:

1. Cohen-Macaulay over a field k if the characteristic of k does not divide $\#G$

2. a pseudomanifold with boundary for all G

3. a pseudomanifold (without boundary) if and only if G contains no reflections

4. an orientable pseudomanifold if and only if G contains no elements of odd length

5. Gorenstein over a field k if and only if either $G = W$, or if both conditions 1 and 4 above hold.

Chapter 3 concerns the theory of P-partitions. It turns out that when W is the symmetric group S_n , and G is among the classes of subgroups G to be considered in Chapter 4, a fundamental domain for the action of G on the Coxeter complex may be readily identified using the language of P-partitions [St3]. The usual theory of P-partitions is an attempt to unify some of the many enumeration results for partitions of a number, partitions into distinct parts, compositions, and tableaux. It starts with a partial order P on numbers $\{1, 2, \ldots, n\}$ (i.e. a *labelled poset*), and defines a P-partition to be a map

$$f : \{1, 2, \ldots, n\} \rightarrow \mathbf{N}$$

satisfying $f(i) \geq f(j)$ whenever $i <_P j$, and $f(i) > f(j)$ whenever $i <_P j$ and $i > j$. Thus a partition into n parts is a P-partition for the partial order

$$1 <_P 2 <_P <_P \ldots <_P n,$$

a partition into n distinct parts is a P-partition for

$$n <_P n - 1 <_P \ldots <_P 1,$$

and a composition into n parts is a P-partition for the partial order P on $\{1, 2, \ldots, n\}$ in which no 2 elements are related. The main result in the usual theory is that the set of $\mathcal{A}(P)$ of all P-partitions decomposes into the disjoint union of all sets $\mathcal{A}(P_\sigma)$ where P_σ is the total order

$$\sigma_1 < \sigma_2 < \ldots < \sigma_n$$

defined by some permutation σ, as σ ranges over the set $\mathcal{L}(P)$ of all permutations which extend P to a total order. For example, if P is the partial order given by $2 <_P 1$ and $2 <_P 3$ on $\{1, 2, 3\}$, then

$$\mathcal{A}(P) = \mathcal{A}(2 < 1 < 3) \amalg \mathcal{A}(2 < 3 < 1)$$

where $\amalg$ denotes disjoint union of sets. This means that

$$\{f \in \mathbf{R}^3 : f(2) > f(1), f(2) \geq f(3)\}$$

$$= \{f \in \mathbf{R}^3 : f(2) > f(1) \geq f(3)\} \amalg \{f \in \mathbf{R}^3 : f(2) \geq f(3) > f(1)\}.$$

In Chapter 3, we show how a partial order on the numbers $\{1, 2, \ldots, n\}$ is equivalent to the choice of a subset of the root system A_{n-1} lying in some pointed cone. We then use this to define the notion of a *parset* (analogous to posets) for other root systems, and proceed to extend some of the theory of P-partitions to this context. In Section 3.2, we show how the decomposition of the root system's ambient space given by the main P-partition result leads to the usual *shelling* of $\Sigma(W, S)$ ([Bj3], [GS]). We also extend the theory of *multipartite P-partitions* ([GG],[Ge2]), and use this to give a shelling of $\Sigma(W^r, rS)$ (where (W^r, rS) is the Coxeter system which is the direct product of r copies of (W, S)). In Section 3.3, we present two immediate applications of the preceding theory:

1. We give a simple proof (suggested by Gessel) of the existence of *Solomon's descent algebra* $\mathcal{S}$ ([So2]) for any finite Coxeter system,

and some lesser known modules over $\mathcal{S}$ recently found by Moszkowski ([Mo]).

2. We prove a generalization to all finite Coxeter groups of the following theorem of Kreweras and Moszkowski: Let ω be a word on letters $\{1, 2, \ldots, n\}$ (with no repeated letters), and $J \subseteq \{1, \ldots, n-1\}$. Then among all permutations of $\{1, 2, \ldots, n\}$ with *descent set J*, the number which contain ω as a subword depends only on the descents of ω.

In Chapter 4, we use this extended P-partition theory to examine quotients by some specific classes of subgroups G. Our general strategy:

1. Identify a fundamental domain, in terms of P-partitions, for the action of G on the ambient vector space of the root system.

2. Decompose this fundamental domain using P-partition theory.

3. Turn this decomposition into a *partitioning* or *shelling* of $\Sigma(W,S)/G$.

We apply this strategy to three classes of subgroups G:

1. Reflection subgroups W', i.e. groups generated by the reflections they contain.

2. Alternating subgroups E' of reflection subgroups, i.e. the elements of even length in a reflection subgroup W'.

3. Diagonal embeddings $\Delta^r(W')$ or $\Delta^r(E')$ of the two classes above in the Coxeter system (W^r, rS).

Our results may be summarized as follows:

1. $\Sigma(W^r, rS)/\Delta^r(W')$ and $\Sigma(W^r, rS)/\Delta^r(E')$ are partitionable for all r.

2. $\Sigma(W^r, rS)/\Delta^r(W')$ is shellable for $r = 1, 2$, but may be non-shellable for $r \geq 3$.

3. $\Sigma(W^r, rS)/\Delta^r(E')$ is shellable for $r = 1$, but may be non-shellable for $r \geq 2$

4. $\Sigma(W^2, 2S)/\Delta^2(W')$ and $\Sigma(W, S)/E'$ are homeomorphic to spheres, and $\Sigma(W, S)/W'$ is homeomorphic to a disk.

These partitionings yield combinatorial interpretations of the *type-selected Möbius invariants* β_J:

$$\beta_{J_1, \ldots, J_r}(\Sigma(W^r, rS)/\Delta^r(W'))$$

$$= \quad \#\{(w_1, \ldots, w_r) : D(w_i) = J_i, I(w_r w_{r-1} \cdots w_1) \cap W' = \emptyset\}$$

$$\beta_{J_1, \ldots, J_r}(\Sigma(W^r, rS)/\Delta^r(E'))$$

$$= \quad \#\{(w_1, \ldots, w_r) : D(w_i) = J_i, I(w_r w_{r-1} \cdots w_1) \cap W' = \emptyset \text{ or } T \cap W'\}$$

where T is the set of all reflections of W, and $I(w)$ is the set of *(left) inversions* of w. When $\Sigma(W, S)/G$ is Gorenstein, the *fine Dehn-Somerville* equations assert that $\beta_J = \beta_{S-J}$ for all $J \subseteq S$, and using our earlier criteria for Gorenstein-ness, we produce non-bijective equalities between some of the cardinalities of sets above. In Section 4.2, we apply our partitionings and shellings to the invariant theory problems mentioned at the beginning of this introduction.

In Chapter 5, we examine the quotients $\Sigma(W, S)/\langle c \rangle$ where $\langle c \rangle$ is the cyclic subgroup generated by a *Coxeter element*. Here, partitionings and shelling are harder to come by, and we concentrate rather on finding relations that hold among the β_J's of $\Sigma(W, S)/\langle c \rangle$. Our main results:

1. $\beta_J = \beta_{\phi(J)}$ for all $J \subseteq S$ whenever ϕ is a diagram automorphism of (W, S).

2. $\beta_J + \beta_{J+s} = \beta_{S-J} + \beta_{S-J-s}$ for all $J \subseteq S - s$ whenever c satisfies a condition called *s-duality*. We show (by enumerating the exceptions) that c is s-dual for almost all finite Coxeter systems (W, S) and $s \in S$.

In Section 5.2, we look at a certain filtration of $\Sigma(W, S)$ and $\Sigma(W, S)/\langle c \rangle$ using the notion of *primitivity*. We then use a result of Gessel (and its analogues for some other Coxeter systems) to partition a large piece of this filtration, and deduce some non-bijective equalities similarly to those in Chapter 4.

Chapter 2

Coxeter complexes and their quotients

2.1 Coxeter complexes

Let (W, S) be a *finite Coxeter system* , i.e. W is a finite group generated by Euclidean reflections acting on an **R**-vector space V of dimension $\#S$, and S is its generating set of *simple reflections* (see [Bro] for an excellent introduction to Coxeter systems; the standard reference is [Bo]). We shall give two definitions of the *Coxeter complex* $\Sigma(W, S)$.

Definition(informal): $\Sigma(W, S)$ is the simplicial complex describing the cell decomposition of the unit sphere in V "cut out" by the reflecting hyperplanes of reflections in W.

Definition(formal): Given $J \subseteq S$, let the *standard parabolic subgroup W_J* be the subgroup of W generated by J, i.e $W = \langle J \rangle$. Then $\Sigma(W, S)$ is the

9

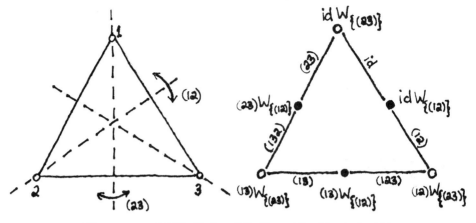

Figure 2.1: $\Sigma(W, S)$ for $(W, S) = (S_3, \{(12), (23)\})$

simplicial complex whose faces are the cosets $\{wW_J\}_{w \in W, J \subseteq S}$ of standard parabolic subgroups, with inclusion of faces corresponding to *reverse inclusion* of cosets (i.e the "face" $w_1 W_{J_1}$ is contained in the "face" $w_2 W_{J_2}$ when $w_2 W_{J_2} \subseteq w_1 W_{J_1}$).

The (non-trivial) facts that both of these define simplicial complexes, and that they are equivalent may be found in [Bro, Chapters 1,3].

Example: Let W be the symmetric group S_n on n letters, and

$$S = \{(12), (23), \ldots, (n-1 \ n)\}$$

the adjacent transpositions. W may be realized as the symmetry group of a regular $(n-1)$-simplex having vertices labelled $\{1, 2, \ldots, n\}$ and centered about the origin in $\mathbf{R}^{n-1}$ (see Fig. 1 for a picture when $n = 3$).

Note that in the figure, $\Sigma(W, S)$ is isomorphic to the barycentric subdivision of the boundary complex of the simplex. It is well-known, and not hard to see that if (W, S) may be realized as the symmetry group of a *regular polytope* $\mathcal{P}$, then $\Sigma(W, S)$ is isomorphic to the *barycentric subdivision* of the

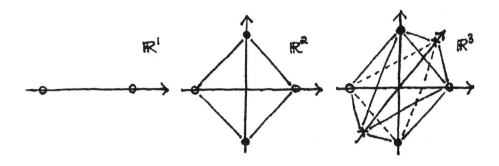

Figure 2.2: $\Sigma(W^r, rS)$ for $W = \mathbf{Z}_2$ and $r = 1, 2, 3$

boundary complex of $\mathcal{P}$. In fact, this is the case exactly when the *Coxeter diagram* of (W, S) (see Section 5.1) is linear.

Example: If (W, S) is a Coxeter system and $r \in \mathbf{P}$, then $\Sigma(W^r, rS)$ is also a Coxeter system, where $W^r = \underbrace{W \times \cdots \times W}_{r \ times}$, and rS is the disjoint union of r copies of S embedded in each coordinate of W^r. It is easy to check (see [Ti, Corollary 2.15]) that

$$\Sigma(W^r, rS) \cong \underbrace{\Sigma(W, S) * \cdots * \Sigma(W, S)}_{r \ times},$$

where $\cong$ denotes isomorphism and $*$ denotes the *join* of simplicial complexes. For example, if $(W, S) = (\mathbf{Z}_2, \{s\})$, then $\Sigma(W, S)$ is just the 0-sphere $\mathbf{S}^0$, and hence

$$\Sigma(W^r, rS) \cong \mathbf{S}^0 * \cdots * \mathbf{S}^0,$$

sometimes known as the *r-dimensional cross-polytope or r-hyperoctahedron* (see Fig. 2 for $r = 1, 2, 3$).

We note two important properties of $\Sigma(W, S)$:

1. $\Sigma(W, S)$ is *(completely) balanced*, i.e. we can label each vertex of $\Sigma(W, S)$ with an element $s \in S$ (call s the *type* of that vertex) so that every maximal face contains exactly one vertex of each type. A vertex of $\Sigma(W, S)$ corresponds to a coset $wW_{S-\{s\}}$ of a maximal proper parabolic subgroup, which we label s. Similarly, we say a face of $\Sigma(W, S)$ corresponding to a coset wW_J is of type $S - J$ (since it lies above one vertex of each type in $S - J$). See [Bro] Chapter 3, or [Ti] Definition 2.5 for more on this labelling.

2. The Coxeter group W acts on $\Sigma(W, S)$ as the group of simplicial automorphisms that preserve the above labelling. In fact, this action is simply the action of W on left cosets, i.e. given $g \in W$ and wW_J a face of $\Sigma(W, S)$, the image of wW_J under the action of g is the face gwW_J (see [Bro] Chapter 3).

2.2   Quotients of Coxeter complexes

Let G be a subgroup of W. Since W acts on $\Sigma(W, S)$, so does G, and we now give two definitions of the quotient $\Sigma(W, S)/G$ of $\Sigma(W, S)$ under this action.

Definition(topological): $\Sigma(W, S)/G$ is the quotient space of the sphere $\Sigma(W, S)$ under the action of G. That is, let $\Sigma(W, S)/G$ as a set be the set of G-orbits of points on the sphere $\Sigma(W, S)$, and give this set the quotient topology induced by the canonical surjection $\pi : \Sigma(W, S) \rightarrow \Sigma(W, S)/G$.

Definition(combinatorial): Let $\Sigma(W, S)/G$ be the poset of orbits of faces of $\Sigma(W, S)$, i.e double cosets $\{GwW_J\}_{w \in W, J \subseteq S}$, ordered under reverse inclusion of double cosets.

Because the action of G is label-preserving, the elements of the poset $\Sigma(W, S)/G$ still have well-defined labels, namely that the double coset GwW_J has label $S - J$. Furthermore, any maximal element of this poset must lie above exactly one element of each type $K \subseteq S$ (since two faces of $\Sigma(W, S)$ of different types cannot be identified by an element of G). Thus, $\Sigma(W, S)/G$ is what is known as a *simplicial poset* [St2] or *Boolean complex* [GS] or *complex of Boolean type* [Bj1], i.e. a poset with a least element $\hat{0}$ in which every lower interval $[\hat{0}, x]$ is isomorphic to a Boolean algebra. Simplicial posets are a generalization of the face posets of simplicial complexes, in that they correspond to the face posets of regular CW-complexes in which each maximal face (with its boundary faces) is combinatorially isomorphic to a simplex (see [Bj1,St2]). It is straightforward to show that the topological definition of $\Sigma(W, S)/G$ may be given such a CW-complex structure so that the combinatorial definition of $\Sigma(W, S)/G$ is its face poset. We call this CW-complex the *topological realization* of the associated simplicial poset, and in what follows we will often not distinguish between the two of them.

Example: Let $(W, S) = (S_3, \{(12), (23)\})$, and let G be the cyclic subgroup generated by the 3-cycle (123). Fig. 3 shows the two ways of viewing $\Sigma(W, S)/G$.

Example: Let $(W, S) = (\mathbf{Z}_2, \{s\})$ and (W^r, rS) be as before, and let G be the *diagonal embedding* of W into W^r, i.e. $G = \langle (s, \ldots, s) \rangle \subseteq W^r$. One can check that the non-trivial element of G acts on the r-hyperoctahedron by

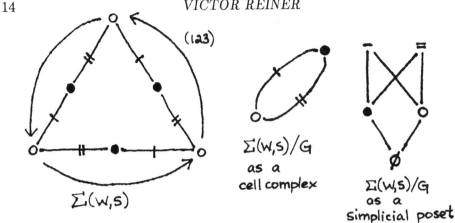

Figure 2.3: $\Sigma(W,S)/G$ for $(W,S) = (S_3, \{(12),(23)\}), G = \langle(123)\rangle$

swapping antipodal vertices, and hence acts as the antipodal map on all of $\Sigma(W^r, rS)$. Hence $\Sigma(W^r, rS)/G$ is homeomorphic to $(r-1)$-dimensional *real projective space* $\mathbf{R}P^{r-1}$.

By analogy to simplicial complexes, we will call the elements of a simplicial poset *faces*, maximal elements *facets*, and minimal non-$\hat{0}$ elements *vertices*. Note that the remarks following Definitions 1 and 2 imply that every facet of $\Sigma(W,S)/G$ lies above exactly one vertex of each type $s \in S$. When the vertices of a simplicial poset P can be labelled in such a fashion, we say P is *(completely) balanced*.

2.3   General results about quotients

We now study the combinatorial and topological properties of $\Sigma(W,S)/G$ via its *face ring*. The *Stanley-Reisner ring* (or *face ring*) $k[\Delta]$ of a simplicial complex Δ has been used extensively in the study of simplicial complexes (see e.g. [St1], Chapter 2). In [St2], Stanley introduced the *face ring* $k[P]$ for a simplicial poset P, which he defined as follows:

Definition: Let k be a field. Then $k[P]$ is the quotient $k[x]_{x \in P}/I_P$ of the polynomial ring in the faces of P by the ideal I_P having the following generators:

1. xy if x and y have no common upper bound in P

2.
$$xy - (x \wedge y) \left(\sum_{z \in \mathrm{mub}\{x,y\}} z \right)$$

   if x, y have a common upper bound b in P. Here $\mathrm{mub}\{x,y\}$ is the set of all *minimal* upper bounds for x, y, and $x \wedge y$ is the greatest lower bound of x, y ($x \wedge y$ exists because x, y both lie in the Boolean algebra $[\hat{0}, b]$).

3. $\hat{0} - 1$ (i.e. the $\hat{0}$ element of P is identified with the unit of the ring $k[P]$).

This definition reduces to the standard face ring $k[\Delta]$ when P is the face poset of a simplicial complex Δ.

We will be dealing exclusively with balanced simplicial posets P having a label-preserving G-action (i.e. G is a subgroup of automorphisms of P satisfying $\mathrm{type}(gx) = \mathrm{type}(x) \; \forall g \in G, x \in P$).

Definition: With P and G as above, the *quotient poset* P/G has as elements the G-orbits $\{Gx\}_{x \in P}$, with $Gx \leq Gy$ if $gx \leq y$ in P for some $g \in G$. The remarks following the combinatorial definition of $\Sigma(W, S)/G$ show that P/G is also a simplicial poset. Note that the combinatorial definition of $\Sigma(W, S)/G$ is the special case of this where P is the poset of faces of $\Sigma(W, S)$.

Our first theorem gives a useful relation between $k[P]$ and $k[P/G]$.

Theorem 2.3.1 *Let P and G be as above, and define a set map $\phi : P/G \to k[P]$ by*

$$\phi(Gx) = \sum_{x' \in Gx} x'.$$

Then ϕ extends to a ring isomorphism $\tilde{\phi} : k[P/G] \to k[P]^G$, where $k[P]^G$ denotes the G-invariant subring of $k[P]$.

Proof: We first establish some notation. Let S be the labelling set for P. Given $J \subseteq K \subseteq S$ and a face $x \in P$ with $\text{type}(x) = K$, let the *restriction* $\text{Res}_J(x)$ of x to J be the unique face under x of type J.

We must check that ϕ extends to a ring homomorphism $\tilde{\phi} : k[P/G] \to k[P]$. We need to show that

$\phi(Gx)\phi(Gy) =$

$$\begin{cases} \phi(Gx \wedge Gy) \sum_{Gz \in mub\{Gx, Gy\}} \phi(Gz) & \text{if } Gx, Gy \text{ have an upper bound in } P/G \\ 0 & \text{else} \end{cases}$$

If we have that

$$0 \neq \phi(Gx)\phi(Gy) = \sum_{\substack{x' \in Gx \\ y' \in Gy}} x'y',$$

then there must exist $x' = g_1 x$ and $y' = g_2 y$ such that $x'y' \neq 0$ in $k[P]$, and hence x', y' have and upper bound $z \in P$. But then Gz would be an upper bound for Gx, Gy in P/G. Hence $\phi(Gx)\phi(Gy) = 0$ if Gx, Gy have no upper bound in P/G.

Otherwise, let $mub\{Gx, Gy\} = \{Gz_i\}_{i=1,\dots,r}$. Without loss of generality, we may pick x, y so that they have an upper bound in P, and hence $x \wedge y$

exists, with $\text{type}(x \wedge y) = \text{type}(x) \cap \text{type}(y)$. Then $Gx \wedge Gy = G(x \wedge y)$, since $G(x \wedge y) \leq Gx, Gy$ and

$$
\begin{aligned}
\text{type}(G(x \wedge y)) &= \text{type}(x \wedge y) \\
&= \text{type}(x) \cap \text{type}(y) \\
&= \text{type}(Gx) \cap \text{type}(Gy) \\
&= \text{type}(Gx \wedge Gy).
\end{aligned}
$$

Thus,

$$
\begin{aligned}
\phi(Gx \wedge Gy) \sum_{Gz \in mub\{Gx,Gy\}} \phi(Gz) &= \left(\sum_{i=1}^{r} \phi(Gz_i) \right) \phi(G(x \wedge y)) \\
&= \left(\sum_{i=1}^{r} \sum_{z \in Gz_i} z \right) \left(\sum_{w \in G(x \wedge y)} w \right) \\
&= \sum_{i=1}^{r} \sum_{z \in Gz_i} \sum_{w \in G(x \wedge y)} zw.
\end{aligned}
$$

Meanwhile,

$$
\begin{aligned}
\phi(Gx)\phi(Gy) &= \sum_{\substack{x' \in Gx \\ y' \in Gy}} x'y' \\
&= \sum_{x' \in Gx, y' \in Gy} x' \wedge y' \left(\sum_{z \in mub\{x',y'\}} z \right) \\
&= \sum_{i=1}^{r} \sum_{z \in Gz_i} z \left(\sum_{\substack{x' \in Gx, y' \in Gy \\ z \in mub\{x',y'\}}} x' \wedge y' \right) \\
&= \sum_{i=1}^{r} \sum_{z \in Gz_i} \sum_{w \in G(x \wedge y)} zw \cdot \#\{x' \in Gx, y' \in Gy : \\
&\qquad\qquad z \in \text{mub}\{x', y'\}, w = x' \wedge y'\}.
\end{aligned}
$$

Therefore it suffices to show $\forall w \in G(x \wedge y), z \in Gz_i$ that

$$
zw = zw \cdot \#\{x' \in Gx, y' \in Gy : z \in \text{mub}\{x', y'\}, w = x' \wedge y'\}.
$$

If z, w have no upper bound in P, then it is trivially true since $zw = 0$. If z, w have an upper bound $v \in P$, then the cardinality of the set on the right is 1, because x', y' are uniquely defined by $x' = \mathrm{Res}_{\mathrm{type}(x)} v$ and $y' = \mathrm{Res}_{\mathrm{type}(y)} v$. So it is still true.

Thus ϕ extends to a ring homomorphism $\tilde{\phi} : k[P/G] \to k[P]$. It is clear that the image of $\tilde{\phi}$ is contained in $k[P]^G$, since

$$\tilde{\phi}(Gx) = \sum_{x' \in Gx} x' \in k[P]^G,$$

and $\{Gx\}_{x \in P}$ generate $k[P/G]$ as an algebra.

It only remains to show that $\tilde{\phi}$ takes a k-basis for $k[P/G]$ to one for $k[P]^G$. From [St2], Lemma 3.4, we know that a k-basis for $k[P]$ consists of all monomials $x_1 x_2 \cdots x_r$ supported on a multichain $x_1 \leq \ldots \leq x_r$ in P. Thus we have a k-basis for P/G consisting of all monomials $Gx_1 Gx_2 \cdots Gx_r$ with $Gx_1 \leq \ldots \leq Gx_r$ in P/G. We also know that we can get a basis for $k[P]^G$ by *symmetrizing* the basis for $k[P]$, i.e. we take all sums of the form $\sum_{g \in G} g(x_1 \cdots x_r)$ with $x_1 \leq \ldots \leq x_r$ in P.

Having identified our two bases, we have

$$\begin{aligned}
\tilde{\phi}(Gx_1 \cdots Gx_r) &= \tilde{\phi}(Gx_1) \cdots \tilde{\phi}(Gx_r) \\
&= \sum_{(g_1, \ldots, g_r) \in G^r} (g_1 x_1) \cdots (g_r x_r).
\end{aligned}$$

Note that if $x_1 \leq x_2$ in P and $(g_1 x_1)(g_2 x_2) \neq 0$ in $k[P]$, then $g_1 x_1, g_2 x_2$ have some upper bound z, and hence

$$\begin{aligned}
g_1 x_1 &= \mathrm{Res}_{\mathrm{type}(x_1)} z \\
&= \mathrm{Res}_{\mathrm{type}(x_1)} (\mathrm{Res}_{\mathrm{type}(x_2)} z)
\end{aligned}$$

$$= g_2 \mathrm{Res}_{\mathrm{type}(x_1)}(\mathrm{Res}_{\mathrm{type}(x_2)} g_2^{-1} z)$$

$$= g_2 \mathrm{Res}_{\mathrm{type}(x_1)} x_2$$

$$= g_2 x_1$$

since $x_1 \leq x_2 \leq g_2^{-1} z$.

Thus $(g_1 x_1) \cdots (g_r x_r) \neq 0$ if and only if we can replace all the g_i's by a single g, i.e. $g_i x_i = g x_i \ \forall i$ and some $g \in G$. Hence

$$\tilde{\phi}(G x_1 \cdots G x_r) = \sum_{(g_1, \ldots, g_r) \in G^r} (g x_1) \cdots (g x_r) = \sum_{g \in G} g(x_1 \cdots x_r).$$

So $\tilde{\phi}$ takes our basis for $k[P/G]$ into our basis for $k[P]^G$, and hence is an isomorphism.∎

Example: Let P be the balanced simplicial poset shown in Fig. 4, with $G = \mathbf{Z}_2$ acting on P by swapping a and b, leaving all other faces fixed. Then P/G is as shown, and we have

$$k[P] = k[a, b, c, d]/(ab, cd - (a + b))$$

$$k[P/G] = k[\tilde{c}, \tilde{d}, \tilde{e}]/(\tilde{c}\tilde{d} - \tilde{e})$$

and it is easy to see that

$$\tilde{\phi} : \tilde{c} \mapsto c, \tilde{d} \mapsto d, \tilde{e} \mapsto a + b$$

is an isomophism from $k[P/G] \to k[P]^G$.

In [St2], Stanley defines a simplicial poset P to be *Cohen-Macaulay over the field k* (abbreviated CM/k) if $k[P]$ satisfies a ring-theoretic condition known as *Cohen-Macaulay-ness*. For face rings, this condition turns out to be equivalent to purely topological conditions on the realization of P, namely

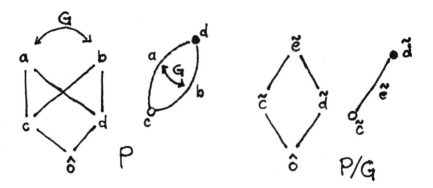

Figure 2.4: An example of P and P/G

that for $i < dim(P)$ we have $\tilde{H}_i(P;k) = \tilde{H}_i(P, P - p) = 0$ for all points p in the realization of P, where $\tilde{H}$ denotes reduced homology (see [St2] for more details).

Theorem 2.3.2 *Let P be a balanced simplicial poset and G a group of label-preserving automorphisms. If P is CM/k, and the characteristic of the field k does not divide $\#G$, then P/G is CM/k.*

Proof: $k[P/G] \cong k[P]^G$ by the previous theorem. Since P is CM/k, we know that $k[P]$ is a Cohen-Macaulay ring. Since the characteristic of k does not divide $\#G$, we can apply a result of Hochster and Eagon ([HE], Proposition 13) to conclude that $k[P]^G$ is also Cohen-Macaulay. Therefore $k[P/G]$ is CM/k. ∎

Corollary 2.3.3 *Let (W, S) be a finite Coxeter system, and G a subgroup of W. Then $\Sigma(W, S)/G$ is CM/k for all fields k whose characteristic does not divide $\#G$.*

Proof: Since $\Sigma(W, S)$ is a sphere, it is CM/k for all fields k (by the topological characterization). Now apply Theorem 2.3.2.∎

Theorem 2.3.2 may also be used to prove a result about simplicial posets which are not necessarily balanced. Let P be a simplicial poset and G a group of automorphisms of P. If we let $|P|$ denote the topological realization of P as a cell complex, then G acts as a group of homeomorphisms of $|P|$ and we may form the quotient space $|P|/G$. Although $|P|/G$ does not carry an obvious cell-structure that would make it the realization of a simplicial poset, we can still speak of $|P|/G$ being CM/k by using the topological characterization.

Theorem 2.3.4 *Let P and G be as in the preceding paragraph. If P is CM/k, and the characteristic of k does not divide $\#G$, then $|P|/G$ is CM/k.*

Proof: Let $Sd(P)$ denote the barycentric subdivision of P, i.e. the simplicial complex of all chains in P. $Sd(P)$ is a balanced simplicial complex in which the label of a vertex is given by the dimension of the face of P to which it corresponds, and hence G acts a label-preserving group of automorphisms of $Sd(P)$. Note also that $|Sd(P)|$ is homeomorphic to $|P|$, and one can easily check that $|Sd(P)/G|$ is homeomorphic to $|P|/G$ (where $Sd(P)/G$ is the quotient simplicial poset of $Sd(P)$ under the action of G). Since P is CM/k, so is $Sd(P)$ (as CM-ness is a topological property). By Theorem 2.3.2, so is $Sd(P)/G$, and hence so is $|P|/G$.∎

This is not the end of the story. One might suspect that there is a purely topological version of the same theorem. In fact, using the main result of [Sm], one can prove the following [1]:

---

[1]Thanks to K. Brown for suggesting a topological approach, H. Miller for the key reference [Sm], and H. Sadofsky for technical help

Theorem 2.3.5 *Let X be a Hausdorff space, G a finite group of homeomorphisms of X, and k a field whose characteristic does not divide $\#G$. Let $\pi : X \to X/G$ be the quotient mapping. Then $\forall i \in \mathbf{N}, x \in X$ we have that $\tilde{H}_i(X/G; k)$ is a direct summand of $\tilde{H}_i(X; k)$ and $\tilde{H}_i(X/G, X/G - \pi(x); k)$ is a direct summand of $\tilde{H}_i(X, X - x; k)$.* ∎

Since the topological characterication of CM/k asserts the vanishing of certain groups $\tilde{H}_i(X; k)$ and $\tilde{H}_i(X, X - x; k)$, we clearly have

$$\text{Theorem 2.3.5} \Rightarrow \text{Theorem 2.3.4} \Rightarrow \text{Theorem 2.3.2.}$$

2.4   Further general results about $\Sigma(W, S)$

We now focus our attention on $\Sigma(W, S)/G$ rather than more general quotients, and investigate some important combinatorial invariants associated to them.

Definition: Given a balanced simplicial poset P with label set S, and $J \subseteq S$, let $\alpha_J(P)$ be the number of faces of P of type J, and let

$$\beta_J(P) = \sum_{K \subseteq J} (-1)^{\#(J-K)} \alpha_K(P).$$

$\beta_J(P)$ is sometimes called the *J-type-selected Möbius invariant* of P, and the α's and β's are sometimes called the *fine f-vector* and *fine h-vector* of P respectively.

A priori, $\beta_J(P) \in \mathbf{Z}$. However, if P is $CM/\mathbf{Q}$ then it is known that $\beta_J(P) \in \mathbf{N}$, and in fact $\beta_J(P)$ has the following two alternate interpretations in this case:

1.

$$\beta_J(P) = dim_{\mathbf{Q}}(\mathbf{Q}[P]/(\theta_s)_{s\in S})_J$$

where $(\theta_s)_{s\in S}$ is the ideal generated by the *rank-row polynomials* $\theta_s = \sum_{\substack{x\in P \\ type(x)=s}} x$ in $k[P]$, and $(\mathbf{Q}[P]/(\theta_s)_{s\in S})_J$ is the Jth graded-homogeneous component of the quotient ring $\mathbf{Q}[P]/(\theta_s)_{s\in S}$. In fact, knowing $\beta_J(P)$ for all $J \subseteq P$ gives an expression for the finely graded *Hilbert series* of $\mathbf{Q}[P]$. See [Ga], Section 2 for more details.

2.

$$\beta_J(P) = dim_{\mathbf{Q}}(\tilde{H}_{\#J-1}(P_J;\mathbf{Q}))$$

where $\tilde{H}_{\#J-1}(P_J;\mathbf{Q})$ denotes the $(\#J - 1)^{st}$ reduced homology group with rational coefficients, and P_J is (the realization of) the simplicial poset obtained from P by deleting those faces $x \in P$ with type(x) not contained in J. P_J is sometimes called the *J-type-selected subcomplex* of P. See [St4], Section 1 for more details.

Since $\Sigma(W, S)/G$ is always $CM/\mathbf{Q}$ (by Corollary 2.3.3), the above facts apply. In light of the second interpretation above, our next result allows us to calculate $dim_{\mathbf{Q}}(\tilde{H}_{\#S-1}(P_J;\mathbf{Q}))$.

Proposition 2.4.1

$$\beta_S(\Sigma(W, S)/G\ ) = \begin{cases} 1 & \text{if } sgn(g) = 1 \text{ for all } g \text{ in } G \\ 0 & \text{else} \end{cases}$$

where sgn denotes the sign character of W, i.e. $sgn(g)$ is the determinant of g thought of as a linear transformation of V (note that $sgn(g) = \pm 1$ since W is generated by reflections).

Proof:

$$
\begin{aligned}
\beta_S(\Sigma(W,S)/G\,) &= \sum_{K \subseteq S} (-1)^{\#(S-K)} \alpha_K(\Sigma(W,S)/G\,) \\
&= \sum_{K \subseteq S} (-1)^{\#(S-K)} \#\{\text{double cosets } GwW_{S-K} \subseteq W\} \\
&= \sum_{K \subseteq S} (-1)^{\#(S-K)} \langle \mathrm{Ind}_G^W 1_G, \mathrm{Ind}_{W_{S-K}}^W 1_{W_{S-K}} \rangle_W
\end{aligned}
$$

by Mackey's formula ([Se], Chapter 7), where here Ind denotes induction of characters, $1_G, 1_{W_{S-K}}$ are the trivial characters of G, W_{S-K} respectively, and $\langle \cdot, \cdot \rangle_W$ is the inner product of characters of W. Thus

$$
\begin{aligned}
\beta_S(\Sigma(W,S)/G\,) &= \langle \mathrm{Ind}_G^W 1_G, \sum_{K \subseteq S} (-1)^{\#(S-K)} \mathrm{Ind}_{W_{S-K}}^W 1_{W_{S-K}} \rangle_W \\
&= \langle \mathrm{Ind}_G^W 1_G, \mathrm{sgn} \rangle_W \\
&= \langle 1_G, \mathrm{Res}_G^W \mathrm{sgn} \rangle_G
\end{aligned}
$$

where the second-to-last equality comes from a result of Solomon ([Sol], Theorem 2), and the last equality is by Frobenius reciprocity (here Res denotes restriction of characters, and $\langle \cdot, \cdot \rangle_G$ is the inner product of characters of G; see [Se], Chapter 7). Thus

$$
\beta_S(\Sigma(W,S)/G\,) = \begin{cases} 1 & \text{if } \mathrm{Res}_G^W \mathrm{sgn} = 1_G \\ 0 & \text{else} \end{cases}
$$

by the orthogonality of irreducible characters of G. This is a rephrasing of our result. ∎

Our next result tells us about the singularities of $\Sigma(W,S)/G$.

Definition: A simplicial poset P is a *pseudomanifold with boundary* if

1. P is pure, i.e. all its facets have the same dimension.

2. Any two facets F, F' of P can be joined by a sequence $F = F_1, F_2, \ldots, F_r = F'$ of facets in which F_i, F_{i+1} share a common face of codimension 1 for $1 \leq i < r$.

3. Every face of codimension 1 lies in at most 2 facets.

We say P is a *pseudomanifold (without boundary)* if every face of codimension 1 lies in *exactly two* facets. If P is a pseudomanifold, we say P is *orientable* if it is possible to choose an orientation on the each of the facets of P (i.e. a ± 1 coefficient on each facet) so as to make the sum of all the facets a homology cycle (see [St1], Chapter 0, Defs. 3.15,3.16 and Chapter 2, Theorem 5.1).

Proposition 2.4.2

1. *$\Sigma(W, S)/G$ is always a pseudomanifold with boundary.*

2. *$\Sigma(W, S)/G$ is a pseudomanifold if and only if G contains no reflections (conjugates of elements of S).*

3. *$\Sigma(W, S)/G$ is an orientable pseudomanifold if and only if $sgn(g) = 1$ $\forall g \in G$.*

Proof:

1. Clearly $\Sigma(W, S)/G$ is pure, and any two facets F, F' in $\Sigma(W, S)/G$ can be joined by a sequence as in the definition; simply lift them to facets

$\tilde{F}, \tilde{F}'$ in $\Sigma(W, S)$, join these facets by such a sequence of facets in $\Sigma(W, S)$ (which exists because $\Sigma(W, S)$ is a sphere and hence a pseudomanifold), and then project this sequence down by $\pi : \Sigma(W, S) \to \Sigma(W, S)/G$. Given a face F of codimension 1, F must correspond to a double coset of the form $GwW_{\{s\}}$ for some $w \in W, s \in S$, and hence F lies in the facet(s) corresponding to $GwW_{\emptyset} = Gw$ and $GwsW_{\emptyset} = Gws$. Thus F lies in two facets if $Gw = Gws$, or one facet if $Gw \neq Gws$.

2. By the discussion in 1, $\Sigma(W, S)/G$ is a pseudomanifold exactly when $Gw \neq Gws \ \forall w \in W, s \in S$. Since $Gw = Gws \Leftrightarrow wsw^{-1} \in G$, the result follows.

3. Clearly $\text{sgn}(g) = 1 \ \forall g \in G$ implies G contains no reflections, and hence that $\Sigma(W, S)/G$ is a psudomanifold by 2. On the other hand, it is easy to see that for any pseudomanifold X of dimension d, we have

$$H_d(X; \mathbf{Q}) = \begin{cases} \mathbf{Q} & \text{if } X \text{ is orientable} \\ 0 & \text{else} \end{cases}$$

Hence by Proposition 2.4.1 and our second interpretation of $\beta_S(\Sigma(W, S)/G)$, the result follows. ∎

Definition: A simplicial poset P which is CM/k and also an orientable pseudomanifold is called *Gorenstein* over k (abbreviated Gor^*/k). Like Cohen-Macaulay-ness, this condition can also be defined as a ring-theoretic condition on $k[P]$ which turns out to be equivalent to the purely topological condition that P is a k-homology sphere. See [St2] for details.

Corollary 2.4.3 *Let k be a field whose characteristic does not divide $\#G$. Then $\Sigma(W, S)/G$ is $Gor^*/k \Leftrightarrow sgn(g) = 1 \ \forall g \in G$.* ∎

Remark: There is a slightly weaker condition on a simplicial poset P than being Gor^*/k, that of being *Gorenstein over k*. In [St2], Section 4, Stanley defines this concept and points out that the only simplicial posets which are Gor/k but not Gor^*/k are the Boolean algebras. Hence Gorenstein-ness is only a trivially weaker notion than Gorenstein*-ness. In our context, it is easy to see that $\Sigma(W, S)/G$ is a Boolean algebra if and only if $G = W$, since this would mean that $\Sigma(W, S)/G$ had only a single facet GwW_0, i.e. $Gw = Gw' \ \forall w, w' \in W$.

Gorenstein* simplicial posets satisfy a duality related to *Alexander duality)* (see [St4], Section 2). This is reflected in the following result.

Proposition 2.4.4 *Let P be a balanced simplicial poset (with label set S), which is also $Gor^*/\mathbf{Q}$. Then the invariants $\beta_J(P)$ satisfy the* fine Dehn-Somerville equations:

$$\beta_J(P) = \beta_{S-J}(P) \ \forall J \subseteq S.$$

Sketch of proof: Let $x_1, \ldots, x_n$ be the vertices of P, thought of as independent indeterminates, and define a generating function

$$L_P(x_1, \ldots, x_n) = \sum_{faces \ x \in P} \prod_{x_i \leq x} \frac{x_i}{1 - x_i}$$

Let $\{t_s\}_{s \in S}$ be a another set of independent indeterminates, one for each element of the label set S, and let T be the map from power series in $x_1, \ldots, x_n$ to power series in $\{t_s\}_{s \in S}$ which sends $x_i \mapsto t_{\text{type}(x_i)}$. Then we have

$$T(L_P(x_1, \ldots, x_n)) = T\left(\sum_{faces \ x \in P} \prod_{x_i \leq x} \frac{x_i}{1 - x_i} \right)$$

$$= \sum_{J \subseteq S} \alpha_J(P) \prod_{s \in J} \frac{t_s}{1 - t_s}$$

$$= \frac{\sum_{J \subseteq S} \beta_J(P) \prod_{s \in J} t_s}{\prod_{s \in S} 1 - t_s}$$

Proposition 4.4 of [St2] says that

$$L_P(x_1, \ldots, x_n) = (-1)^{\#S} L_P\left(\frac{1}{x_1}, \ldots, \frac{1}{x_n}\right).$$

Applying the map T gives

$$\frac{\sum_{J \subseteq S} \beta_J(P) \prod_{s \in J} t_s}{\prod_{s \in S} 1 - t_s} = (-1)^{\#S} \frac{\sum_{J \subseteq S} \beta_J(P) \prod_{s \in J} \frac{1}{t_s}}{\prod_{s \in S} 1 - \frac{1}{t_s}}$$

which (with a little algebra) implies our result. $\blacksquare$

Corollary 2.4.5 *If* $sgn(g) = 1 \ \forall g \in G$ *then* $\beta_J(\Sigma(W,S)/G) = \beta_{S-J}(\Sigma(W,S)/$
$\forall J \subseteq S$.

In Chapters 3 and 4, we will give combinatorial interpretations of these non-negative integers $\beta_J(\Sigma(W,S)/G)$ for certain groups G, and then use this corollary to assert non-trivial equalities between cardinalities of certain sets.

Example: Let $(W,S) = (\mathbf{Z}_2, \{s\})$ and (W^r, rS) be as before, and let $G = \langle (s, \ldots, s) \rangle \subseteq W^r$ as before. Then the quotient $\Sigma(W,S)/G \cong \mathbf{R}P^{r-1}$ is CM/k whenever the characteristic of k is not 2. Since $(s, \ldots, s)$ is not a reflection (unless $r = 1$), and $sgn((s, \ldots, s)) = (-1)^r$ we conclude that $\Sigma(W,S)/G$ is a pseudomanifold $\forall r \geq 2$, and orientable $\forall r$ even. Of course, these facts agree with what is known about $\mathbf{R}P^{r-1}$.

Example: Let $(W, S) = (S_3, \{(12), (23)\})$ and $G = \langle (123) \rangle$ as before. Since $\text{sgn}((123)) = 1$, $\Sigma(W, S)/G$ is an orientable pseudomanifold of dimension 1, i.e. $\mathbf{S}^1$ (as shown in Fig. 3). We can use Fig. 3 to write down $\alpha_J(\Sigma(W, S)/G)\ \forall J \subseteq S$, and then calculate β_J from this. This yields the following table:

J	$\alpha_J(\Sigma(W, S)/G)$	$\beta_J(\Sigma(W, S)/G)$
$\emptyset$	1	1
$\{(12)\}$	1	0
$\{(23)\}$	1	0
$\{(12), (23)\}$	2	1

Note that $\beta_J(\Sigma(W, S)/G) = \beta_{S-J}(\Sigma(W, S)/G)\ \forall J \subseteq S$.

Chapter 3

P-partitions for other

Coxeter groups

3.1 Definitions

In this chapter, we generalize some of the theory of P-partitions (see [St3])
which deals with the symmetric group S_n to other finite Coxeter groups. We
will then use some of these results in Chapters 4 and 5 to prove results about
$\Sigma(W,S)/G$ for some specific classes of subgroups G. However, this theory
has some interest on its own, and we present two applications of it in Section
3.3.

Since many of the results of this chapter are known for the case of $W =
S_n$ (see the Introduction), we will try to "translate" the more general results
into these more familiar surroundings whenever possible.

Let (W, S) be a finite Coxeter system acting as a group generated by
reflections on a Euclidean space $(V, \langle \cdot, \cdot \rangle)$, with $dim_{\mathbf{R}} V = \#S$. Let T denote

the reflections of W, i.e. the set of all conjugates in W of elements of S.

Definition: A *positive root system* realizing (W, S) is a pair (Φ, Π) of finite subsets of vectors in V satisfying

1. Π is a basis for V.

2. $\Phi = \Phi^+ \coprod -\Phi^+$ where

$$\Phi^+ = \left\{ \sum c_\alpha \alpha : c_\alpha \in \mathbf{R}, c_\alpha > 0 \right\} \cap \Phi$$

    is the set of all vectors in Φ which can be written as a positive linear combination of vectors in Π,

$$-\Phi^+ = \{-\alpha : \alpha \in \Phi^+\}$$

    and $\coprod$ denotes disjoint union.

3. $S = \{r_\alpha : \alpha \in \Pi\}$ where r_α denotes the reflection through the hyperplane orthogonal to α.

4. $\Phi = W\Pi = \{w\alpha : w \in W, \alpha \in \Pi\}$.

Φ is called the set of *roots*, Φ^+ the *positive roots*, $-\Phi^+$ the *negative roots*, and Π the *simple roots*. What we call here a positive root system realizing (W, S) corresponds in the literature to a *root system* realizing W along with a choice of positive roots consistent with S. See [Bo], Chapitre VI Section 1 for background, and [Bro], Chapter II, Section 5 for a method of constructing (Φ, Π) given any finite Coxeter system (W, S).

Example: Let $(W, S) = (S_n \,, \{(12), (23), \ldots, (n-1 \; n)\})$ acting on

$$V = \left\{ (x_1, \ldots, x_n) \in \mathbf{R}^n : \sum x_i = 0 \right\}$$

by permuting coordinates. Let

$$\Phi = \{e_i - e_j : 1 \le i, j \le n, i \ne j\}$$

where e_i denotes the ith standard basis vector. Let

$$\Phi^+ = \{e_i - e_j : 1 \le i < j \le n\}$$

$$\Pi = \{e_i - e_{i+1} : 1 \le i < n\}.$$

It is easy to check that (Φ, Π) give a positive root system realizing (W, S), which we will call the *standard realization* of S_n . Whenever we say $W = S_n$, we are referring to this realization.

For the remainder of this section, (W, S) will be a finite Coxeter system, and (Φ, Π) a positive root system realizing (W, S).

Definition: A *parset (partial root system)* is a subset $P \subseteq \Phi$ satisfying

1. $\alpha \in P \Rightarrow -\alpha \notin P$

2. If $\alpha_1, \alpha_2 \in P$ and $c_1\alpha_1 + c_2\alpha_2 \in \Phi$ for some $c_1, c_2 > 0$, then $c_1\alpha_1 + c_2\alpha_2 \in P$

The second condition says that P is closed under the operation of taking positive linear combinations that still lie in Φ. We will denote this closure by $\overline{\cdot}^{PLC}$, i.e given $A \subseteq \Phi$ we let $\overline{A}^{PLC}$ be the smallest subset of Φ which is closed under this operation.

Another way of phrasing conditions 1 and 2 is to say that P is the intersection of Φ with some *pointed cone* in V.

Example: For $W = S_n$, a parset P corresponds to a *labelled poset* on

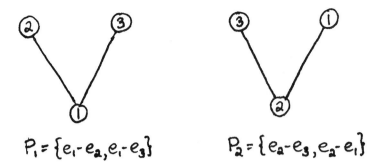

$$P_1 = \{e_1 - e_2, e_1 - e_3\} \qquad P_2 = \{e_2 - e_3, e_2 - e_1\}$$

Figure 3.1: Some parsets for $W = S_n$

the numbers $1, 2, \ldots, n$ (i.e. a partial order $\leq_P$ on $\{1, 2, \ldots, n\}$) via the identification

$$i <_P j \Leftrightarrow e_i - e_j \in P.$$

Conditions 1 and 2 for a parset correspond to *antisymmetry* and *transitivity* of partial orders (reflexivity is already built-in).

Definition: We say 2 parsets P_1, P_2 are *isomorphic* (written $P_1 \cong P_2$) if

$$\exists w \in W \text{ such that } wP_1 = P_2.$$

We say P is *natural* if $P \subseteq \Phi^+$.

Example: For $W = S_n$, two parsets P_1, P_2 are isomorphic if their underlying partial orders (ignoring labels) are isomorphic, i.e.

$$\exists \phi : P_1 \to P_2 \text{ such that } i <_{P_1} j \Leftrightarrow \phi(i) <_{P_2} \phi(j).$$

In Figure 1, $P_1 \cong P_2$, and P_1 is natural, but P_2 is not.

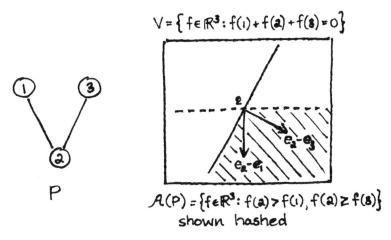

Figure 3.2: An example of $\mathcal{A}(P)$

Definition: A vector $f \in V$ is a *P-partition* if

$$\langle \alpha, f \rangle \geq 0 \,\, \forall \alpha \in P \text{ and}$$

$$\langle \alpha, f \rangle > 0 \,\, \forall \alpha \in P \cap -\Phi^+$$

We denote by $\mathcal{A}(P)$ the set of all P-partitions.

Example: For $W = S_n$, a P-partition is a vector $f = (f(1), \ldots, f(n)) \in \mathbf{R}^n$ satisfying

$$f(i) \geq f(j) \text{ if } i <_P j \text{ and } f(i) > f(j) \text{ if } i <_P j \text{ and } i > j$$

along with the extra condition $\sum f(i) = 0$. This extra condition makes our notion slightly different (in an inessential way) from the usual one given in [St3]. An example of $\mathcal{A}(P)$ for a particular P is shown in Figure 2.

Definition: The *Jordan-Hölder set* of P, denoted $\mathcal{L}(P)$, is the set

$$\{w \in W : P \subseteq w\Phi^+\}.$$

Note that $w\Phi^+$ is a parset, and we say $f \in V$ is *w-compatible* if $f \in \mathcal{A}(w\Phi^+)$.

Given $w \in W$, we define its *(left) inversion set* $I(w)$ and its *(right) descent set* $D(w)$ by

$$I(w) = \Phi^+ \cap w(-\Phi^+)$$

$$D(w) = \Pi \cap w^{-1}(-\Phi^+) = \Pi \cap I(w^{-1})$$

We will also think of $I(w)$ as the subset $\{r_\alpha : \alpha \in I(w)\}$ of T, and $D(w)$ as the subset $\{r_\alpha : \alpha \in D(w)\}$ of S, i.e. we identify a positive root α with the reflection r_α through the hyperplane orthogonal to α. We hope that it will be clear from context which we mean.

Note that since $\Phi^+ = \overline{\Pi}^{PLC}$, we have that

f is w-compatible $\quad\Leftrightarrow\quad \langle \alpha, f \rangle \geq 0 \; \forall \alpha \in w\Phi^+$ and

$$\langle \alpha, f \rangle > 0 \; \forall \alpha \in w\Phi^+ \cap -\Phi^+$$

$$\Leftrightarrow \quad \langle \alpha, w^{-1}f \rangle \geq 0 \; \forall \alpha \in \Phi^+ \text{ and}$$

$$\langle \alpha, w^{-1}f \rangle > 0 \; \forall \alpha \in \Phi^+ \cap w^{-1}(-\Phi^+) = I(w^{-1})$$

$$\Leftrightarrow \quad \langle \alpha, w^{-1}f \rangle \geq 0 \; \forall \alpha \in \Pi \text{ and}$$

$$\langle \alpha, w^{-1}f \rangle > 0 \; \forall \alpha \in \Pi \cap w^{-1}(-\Phi^+) = D(w)$$

In other words, in order to check if f is w-compatible, we only need to look at $\langle \alpha, w^{-1}f \rangle \; \forall \alpha \in \Pi$, rather than looking at $\langle \alpha, f \rangle \; \forall \alpha \in w\Phi^+$

Example: For $W = S_n$, $\mathcal{L}(P)$ is the set of permutations $\sigma = \left(\begin{smallmatrix} 1 & \cdots & n \\ \sigma_1 & & \sigma_n \end{smallmatrix}\right)$ such that the total order $\sigma_1 < \ldots < \sigma_n$ is an extension of P. We have

$$I(\sigma) = \{(ij) : 1 \leq i < j \leq n \text{ and } \sigma^{-1}(i) < \sigma^{-1}(j)\}$$

and

$$D(\sigma) = \{(i \ i+1) : 1 \le i < n \text{ and } \sigma_i > \sigma_{i+1}\}.$$

We also have that f is σ-compatible exactly when f is a P-partition for the total order given by

$$\sigma_1 < \sigma_2 < \cdots < \sigma_n,$$

i.e. when we have $f(\sigma_1) \ge \ldots \ge f(\sigma_n)$ and $f(\sigma_i) > f(\sigma_{i+1})$ for $(i \ i+1) \in D(\sigma)$.

For example, if P is the parset in Figure 2, then $\mathcal{L}(P) = \left\{\binom{123}{213}, \binom{123}{231}\right\}$. If $\sigma = \binom{123}{213}$, then $I(\sigma) = D(\sigma) = \{(12)\}$, and f is σ-compatible if

$$f(2) > f(1) \ge f(3).$$

We come now to the first (and central) result about P-partitions.

Proposition 3.1.1

$$\mathcal{A}(P) \ = \ \coprod_{w \in \mathcal{L}(P)} \mathcal{A}(w\Phi^+)$$

Proof (cf. [Ge2], Theorem 1): We use induction on

$$t = \#\{\alpha \in \Phi^+ : \alpha \notin P, -\alpha \notin P\}.$$

Case 1: $t = 0$. We want to show that $\mathcal{A}(P) = \mathcal{A}(w\Phi^+)$ for some $w \in \mathcal{L}(P)$, so it would suffice to show that $P = w\Phi^+$ for some $w \in W$. Since $t = 0$ implies $\Phi = P \coprod -P$, and $P = \overline{P}^{PLC}$, we conclude that P forms an alternative set of positive roots for Φ (this is essentially the content of [Bo], Chapitre VI, Section 1, No. 7). Since W acts transitively on all possible sets of positive roots, we have $P = w\Phi^+$ for some $w \in W$.

Case 2: $t > 0$. Assume $\alpha, -\alpha \notin P$, and let $P_\alpha = \overline{P \cup \{\alpha\}}^{PLC}$. We claim P_α is a parset, i.e. it also satisfies the first condition for being a parset. To see this, suppose not, i.e. let $\beta, -\beta \in P_\alpha$. Then we must have

$$\beta = a\alpha + \sum a_i \alpha_i$$

$$-\beta = b\alpha + \sum b_i \alpha_i$$

for some $a_i, b_i \geq 0$, $a, b > 0$, and $\alpha_i \in P$. Adding these equations, and dividing by $a + b$ yields

$$-\alpha = \sum \frac{1}{a+b}(a_i + b_i)\alpha_i$$

and hence $-\alpha \in P$, a contradiction. Similarly we can form the parset $P_{-\alpha}$. We then have

$$\mathcal{A}(P) = \mathcal{A}(P_\alpha) \amalg \mathcal{A}(P_{-\alpha})$$

$$\mathcal{L}(P) = \mathcal{L}(P_\alpha) \amalg \mathcal{L}(P_{-\alpha}) .$$

The first equality holds because any $f \in \mathcal{A}(P)$ either satisfies $\langle \alpha, f \rangle \geq 0$ or $\langle -\alpha, f \rangle > 0$. The second equality holds because any $w \in \mathcal{L}(P)$ either satisfies $\alpha \in w\Phi^+$ or $-\alpha \in w\Phi^+$. Thus by induction on t, we are done.∎

An example is shown in Figure 3.

3.2   P-partitions and $\Sigma(W, S)$

We come now to the main link (Lemma 3.2.1) between P-partitions and Coxeter complexes. The theorems of this section are known (see [Bj3], Section 2, [GS], Sections 7,8), however our method of proof is slightly different,

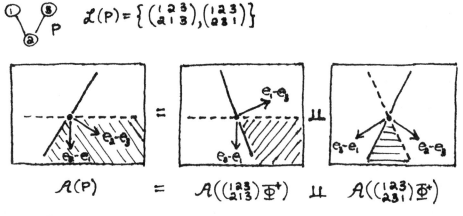

$$\mathcal{L}(P) = \left\{ \begin{pmatrix} 1 & 2 & 3 \\ 2 & 1 & 3 \end{pmatrix}, \begin{pmatrix} 1 & 2 & 3 \\ 2 & 3 & 1 \end{pmatrix} \right\}$$

Figure 3.3: An example of the central result on P-partitions

and will form the prototype for the results in Section 3.4 and Chapters 4 and 5.

Definition: The *fundamental (Weyl) chamber* $\mathcal{C}$ is the set $\mathcal{A}(\Phi^+) \subseteq V$, that is all vectors $f \in V$ satisfying $\langle \alpha, f \rangle \geq 0 \; \forall \alpha \in \Phi^+$ (or alternatively, $\langle \alpha, f \rangle \geq 0 \; \forall \alpha \in \Pi$). Given $f \in V$, let

$$F(f) = \{ w \in W : w^{-1}(f) \in \mathcal{C} \}.$$

We note two important facts about $F(f)$ and $\mathcal{C}$ (see [Bro], Chapter I, Theorem 5F):

1. For $f \in \mathcal{C}$, $F(f) = W_J$ where $J = \{ r_\alpha : \alpha \in \Pi, \langle \alpha, f \rangle = 0 \}$

2. Every $f \in V$ has a unique translate $w(f) \in \mathcal{C}$.

Hence, in general we know that $F(f) = wW_J$ for some $w \in W$ and $J \subseteq S$, i.e. $F(f)$ *always corresponds to a face of* $\Sigma(W, S)$.

Example: For $W = S_n$, $\mathcal{C}$ is the set of all f satisfying $f(1) \geq \ldots \geq f(n)$. The first fact above says that if $f(i) = f(i+1)$, then we can permute the coordinates $i, i+1$ and f will remain in $\mathcal{C}$. E.g. if $f = (3,2,2,1,1,1) \in \mathcal{C}$ then so is $w(f) \in \mathcal{C}$ whenever $w \in W_{\{(23),(45),(56)\}}$. The second fact above says that there is a unique permutation of the coordinates of f into (weakly) decreasing order.

The next lemma establishes the fundamental link between P-partitions and the Coxeter complex $\Sigma(W, S)$, and will be used frequently in our analysis.

Lemma 3.2.1 *For all $f \in V$, we have*

$$f \in \mathcal{A}(w\Phi^+) \ \Leftrightarrow \ wW_\emptyset \subseteq F(f) \subseteq wW_{S-D(w)}$$

and hence for all $g \in W$, we have

$$g(f) \in \mathcal{A}(w\Phi^+) \ \Leftrightarrow \ g^{-1}wW_\emptyset \subseteq F(f) \subseteq g^{-1}wW_{S-D(w)}$$

Proof: We have

$$
\begin{aligned}
f \in \mathcal{A}(w\Phi^+) \quad &\Leftrightarrow \quad \langle \alpha, w^{-1}(f) \rangle \geq 0 \ \forall \alpha \in \Pi \text{ and } \langle \alpha, w^{-1}(f) \rangle > 0 \ \forall \alpha \in D(w) \\
&\Leftrightarrow \quad w \in F(f) \text{ and } \langle -\alpha, w^{-1}f \rangle < 0 \ \forall \alpha \in D(w) \\
&\Leftrightarrow \quad w \in F(f) \text{ and } \langle r_\alpha(\alpha), w^{-1}(f) \rangle < 0 \ \forall \alpha \in D(w) \\
&\Leftrightarrow \quad w \in F(f) \text{ and } \langle \alpha, r_\alpha w^{-1}(f) \rangle < 0 \ \forall \alpha \in D(w) \\
&\Leftrightarrow \quad w \in F(f) \text{ but } wr_\alpha \notin F(f) \ \forall \alpha \in D(w) \\
&\Leftrightarrow \quad wW_\emptyset \subseteq F(f) \subseteq wW_{S-D(w)}.
\end{aligned}
$$

This proves the first assertion. The second follows from the first along with the simple observation that $F(g(f)) = gF(f)$.∎

The previous lemma reflects the following fact, which is tedious but straightforward to verify. Given a face $F = wW_J$ of $\Sigma(W,S)$, define the set

$$V(F) = \{f \in V : F(f) = F\}.$$

Then $V(F) \cap S^{\#S-1}$ is exactly the open cell in the decomposition of the unit sphere $S^{\#S-1}$ corresponding to F (from the "informal" definition of $\Sigma(W,S)$). In fact, our philosophy is to "think of" the face $F = wW_J$ as the same as the open cell $V(F)$. We then analyze the W-action on V (i.e. find a fundamental domain in V for the action of G), and use the previous lemma to translate this into a statement about $\Sigma(W,S)/G$.

Our next theorem exemplifies this philosophy. But first, a definition.

Definition: A *partitioning* or *ER-decomposition* of a simplicial poset P is an expression

$$P = \coprod_{i=1}^{t}[F_i, M_i]$$

where for each i, M_i is a facet of P, F_i is a face of M_i, and

$$[F_i, M_i] = \{F \in P : F_i \le F \le M_i\}.$$

Theorem 3.2.2 $\Sigma(W,S)$ *is partitionable as*

$$\Sigma(W,S) \;=\; \coprod_{w \in W} [wW_{S-D(w)}, wW_\emptyset]$$

Proof: Applying Proposition 3.1.1 to the empty parset $P = \emptyset$, we get

$$V \;=\; \mathcal{A}(\emptyset)$$

$$= \coprod_{w \in \mathcal{L}(\emptyset)} \mathcal{A}(w\Phi^+)$$

$$= \coprod_{w \in W} \mathcal{A}(w\Phi^+)$$

If we now apply the operation $f \mapsto F(f)$ to both ends of the above equation, we get

$$\bigcup_{f \in V} F(f) = \bigcup_{w \in W} \bigcup_{f \in \mathcal{A}(w\Phi^+)} F(f)$$

In light of Lemma 3.2.1 (and the easy-to-check fact that every face $F \in \Sigma(W, S)$ is $F(f)$ for some $f \in V$), this gives

$$\Sigma(W, S) = \coprod_{w \in W} [wW_{S-D(w)}, wW_\emptyset]$$

as we wanted.■

One by-product of a partitioning for a balanced simplicial poset P is another interpretation of $\beta_J(P)$.

Proposition 3.2.3 *If* $P = \coprod_{i=1}^t [F_i, M_i]$ *is a partitioning, then*

$$\beta_J(P) = \#\{i : type(F_i) = J\}$$

Proof: Given $J \subseteq P$, there is exactly one face of type J in each interval $[F_i, M_i]$ with $type(F_i) \subseteq J$. Thus we have

$$\alpha_J(P) = \#\{i : type(F_i) \subseteq J\}$$

Since

$$\beta_J(P) = \sum_{K \subseteq J} (-1)^{\#(J-K)} \alpha_K(P)$$

the result then follows by inclusion-exclusion.■

Corollary 3.2.4

$$\beta_J(\Sigma(W,S)\,) = \#\{w \in W : D(w) = J\}.\blacksquare$$

An important subclass of the partitionable simplicial posets are those that are *shellable*.

Definition: A *shelling* of a simplicial poset P is a partitioning $P = \coprod_{i=1}^{t}[F_i, M_i]$ with the extra condition that $F_i \leq M_j \Rightarrow i \leq j$ (see [Bj3], Proposition 1.2 for the equivalence of this to other definitions of shellings).

Shellability of P has strong consequences for the topology of P and ring theory of $k[P]$. In particular, if P is shellable, then it is CM/k for *all* fields k (see [Bj2] for more on shellability).

In order to shell $\Sigma(W,S)$, we require a bit more technology.

Definition: The *length* of an element $w \in W$ is defined by

$$l(w) = min\{r : w = s_1 s_2 \cdots s_r \text{ for some } s_i \in S\}.$$

The *(right) weak order* $<_R$ is defined to be the transitive closure of these relations:

$$w <_R ws \text{ if } w \in W, s \in S \text{ and } l(w) < l(ws).$$

We note some well-known facts about l and $<_R$:

1. $I(w) = \{t \in T : l(tw) < l(w)\}$ and hence $D(w) = \{s \in S : l(ws) < l(w)\}$ (see [Bo], Chapitre IV, Section 1, Remark after Lemme 3).

2. $l(w) = \#I(w)$ (see [Bo], Chapitre IV, Section 1, Lemme 2).

3. Given wW_J, there is a unique element denoted $\pi^J(w)$ in the coset wW_J satisfying $D(\pi^J(w)) \subseteq S - J$. We also have $\pi^J(w) \leq_R u \ \forall u \in wW_J$ (see [Bo], Chapitre IV, Section 1, Exercice 3, or [Bj3], Introduction to Section 2).

Example: For $W = S_n$,

$$l(\sigma) = \#\{(i,j) : 1 \leq i < j \leq n, \sigma_i > \sigma_j\}$$

(the number of *inversions* of σ). We have $\sigma \leq_R \tau$ if one can get from $\sigma = (\begin{smallmatrix} 1 & \cdots & n \\ \sigma_1 & & \sigma_n \end{smallmatrix})$ to τ by a sequence of exchanges of σ_i, σ_{i+1} with $\sigma_i < \sigma_{i+1}$. Given σ and J, $\pi^J(\sigma)$ is the permutation obtained from σ by arranging the σ_i to be ascending in the places permuted by W_J. For example, letting $\tau = (\begin{smallmatrix} 1234 \\ 3142 \end{smallmatrix})$ we have $l(\tau) = 3$, and for $J = \{(12),(34)\}$ we have $\pi^J(\tau) = (\begin{smallmatrix} 1234 \\ 1324 \end{smallmatrix})$.

Theorem 3.2.5 ([Bj3], Theorem 2.1, [GS], Theorem 8.6)

$$\Sigma(W,S) = \coprod_{w \in W} [wW_{S-D(w)}, wW_\emptyset]$$

is a shelling if we order $\{wW_\emptyset\}_{w \in W}$ *by any linear extension of* $<_R$.

Proof: Let $w_1, w_2, \ldots, w_t$ be such an order. Since Theorem 3.2.2 already asserts that we have a partitioning, we only need to show that

$$w_i W_{S-D(w_i)} \leq w_j W_\emptyset \Rightarrow i \leq j$$

But we have

$$\begin{aligned} w_i W_{S-D(w_i)} \leq w_j W_\emptyset \quad &\Leftrightarrow \quad w_j \in w_i W_{S-D(w_i)} \\ &\Leftrightarrow \quad \pi^{S-D(w_i)}(w_j) = w_i \\ &\Rightarrow \quad w_i \leq_R w_j \\ &\Rightarrow \quad i \leq j \end{aligned}$$

as desired. ∎

3.3 Two applications

In this section we explore two immediate applications of the theory of P-partitions that are only indirectly related to the quotients $\Sigma(W,S)/G$. But first we need to discuss one more object Σ_P associated to a parset P.

Definition: Let (W,S) be finite Coxeter system, and P a (W,S)-parset. We define the subposet $\Sigma_P \subseteq \Sigma(W,S)$ by

$$\Sigma_P = \{F \in \Sigma(W,S) \; : F = F(f) \text{ for some } f \in \mathcal{A}(P) \}$$

Repeating the proof of Theorem 3.2.2 with P in place of the empty parset $\emptyset$ immediately yields

Proposition 3.3.1

$$\Sigma_P = \coprod_{w \in \mathcal{L}(P)} [wW_{S-D(w)}, wW_\emptyset].\blacksquare$$

Although Σ_P is only a subposet of $\Sigma(W,S)$ and not necessarily a simplicial poset, we may still define

$$\alpha_J(\Sigma_P) = \#\{F \in \Sigma(W,S) \; : \text{type}(F) = J\}$$

and

$$\beta_J(\Sigma_P) = \sum_{K \subseteq J} (-1)^{\#(J-K)} \alpha_J(\Sigma_P)$$

as before. As in the preceding section, we conclude

Corollary 3.3.2

$$\beta_J(\Sigma_P) = \#\{w \in \mathcal{L}(P) \; : D(w) = J\}.\blacksquare$$

The following observation is the key to both applications.

Proposition 3.3.3 *Let* $P_i = \overline{A_i}^{PLC}$ *for* $i = 1, 2$ *be two parsets, and suppose* $wA_1 = A_2$ *and* $w(A_1 \cap -\Phi^+) = A_2 \cap -\Phi^+$ *for some* $w \in W$. *Then*

1. $w\mathcal{A}(P_1) = \mathcal{A}(P_2)$ *and*

2. $\beta_J(\Sigma_P) = \beta_J(\Sigma_{P'}) \; \forall J \subseteq S$

Proof: To prove 1, we claim that for $i = 1, 2$ we have

$$f \in \mathcal{A}(P_i) \; \Leftrightarrow \; \langle \alpha, f \rangle \geq 0 \; \forall \alpha \in A_i \text{ and } \langle \alpha, f \rangle > 0 \; \forall \alpha \in A_i \cap -\Phi^+.$$

To see this note that the right implication is obvious, and the only non-trivial part of the left implication is checking that the left side implies

$$\langle \alpha, f \rangle > 0 \; \forall \alpha \in P_i \cap -\Phi^+.$$

To see this, assume $\alpha \in P_i \cap -\Phi^+$, so that we can write $\alpha = \sum c_j \beta_j$ for some $c_j > 0$ and $\beta_j \in A_i$. Now if $\beta_j \in \Phi^+ \; \forall j$, then we reach the contradiction that $\alpha \in \Phi^+$. Hence $\beta_j \in A_i \cap -\Phi^+$ for some j_0, and thus

$$\langle \alpha, f \rangle = c_{j_0} \langle \beta_{j_0}, f \rangle + \sum_{j \neq j_0} c_j \langle \beta_j, f \rangle \; > 0$$

Given this claim, the fact that $w\mathcal{A}(P_1) = \mathcal{A}(P_2)$ follows directly from our hypotheses. To prove 2, we deduce from 1 that $w\Sigma_{P_1} = \Sigma_{P_2}$, and since the action of w is type- preserving, that

$$\alpha_J(\Sigma_{P_1}) = \alpha_J(\Sigma_{P_2}) \; \forall J \subseteq S$$

Then 2 follows immediately.∎

Our first application is a result of Moszkowski, which generalizes a result of Solomon.

Theorem 3.3.4 *Let $J' \subseteq J \subseteq \Pi$ and $K' \subseteq K \subseteq \Phi$. Then for a given $w \in W$,*

$$\#\{(u,v) \in W^2 : D(u) \cap J = J', I(v^{-1}) \cap K = K', \text{ and } uv = w\}$$

depends only on $I(w^{-1}) \cap K$.

Proof: Given $w \in W$, define a parset $P_w = \overline{w(K - K')}^{PLC}$ and notice that

$$
\begin{aligned}
\beta_J(\Sigma_{P_w}) &= \#\{u \in \mathcal{L}(P_w) : D(u) = J\} \\
&= \#\{u \in W : P_w \subseteq u\Phi^+, D(u) = J\} \\
&= \#\{u \in W : w(K - K') \subseteq u\Phi^+, D(u) = J\} \\
&= \#\{u \in W : u^{-1}w(K - K') \subseteq \Phi^+, D(u) = J\} \\
&= \#\{u \in W : I(u^{-1}w) \cap K \subseteq K', D(u) = J\} \\
&= \#\{(u,v) \in W^2 : I(v^{-1}) \cap K \subseteq K', D(u) = J, \text{ and } uv = w\}
\end{aligned}
$$

Thus, by inclusion-exclusion on the set K, it would suffice to show that $\beta_J(\Sigma_{P_w})$ depends only on $I(w^{-1}) \cap K$. So suppose $w, w' \in W$ satisfy $I(w^{-1}) \cap K = I(w'^{-1}) \cap K$. We can apply the previous proposition, once we note that $w'w^{-1} \cdot w(K - K') = w'(K - K')$ and

$$
\begin{aligned}
w'w^{-1} \cdot (w(K - K') \cap -\Phi^+) &= w'((K - K') \cap I(w^{-1})) \\
&= w'((K - K') \cap I(w'^{-1})) \\
&= w'(K - K') \cap -\Phi^+
\end{aligned}
$$

where the second equality follows from our supposition.∎

Corollary 3.3.5 ([Mo], Théorème 1, cf. [So2], Theorem 1) *Given $K' \subseteq K \subseteq \Phi^+$, let $X_{K'}^K$ denote the formal sum*

$$X_{K'}^K = \sum_{\substack{w \in W \\ I(w^{-1}) \cap K = K'}} w$$

as an element of the group ring $\mathbf{Z}W$ of W. Then

1. *$\forall J \subseteq \Pi$, $\{X_{J'}^J\}_{J' \subseteq J}$ span a subring $\mathcal{S}_J$ of $\mathbf{Z}W$ ($\mathcal{S}_\Pi$ is actually a ring with unit and is sometimes called the* Solomon algebra *or* descent algebra *of W).*

2. *$\forall K \subseteq \Phi^+$, $\{X_{K'}^K\}_{K' \subseteq K}$ span an $\mathcal{S}_J$-submodule of $\mathbf{Z}W$ for each $J \subseteq \Pi$.*

Proof: We only need to show that $\forall J' \subseteq J \subseteq \Pi$, and $K' \subseteq K \subseteq \Phi^+$, we have that $X_{J'}^J X_{K'}^K$ is in the $\mathbf{Z}$-span of $\{X_{K''}^K\}_{K'' \subseteq K}$. We have

$$X_{J'}^J X_{K'}^K$$

$$= \sum_{\substack{(u,v) \in W^2 \\ D(u) \cap J = J', I(v^{-1}) \cap K = K'}} uv$$

$$= \sum_{w \in W} w \cdot \#\{(u,v) \in W^2 : D(u) \cap J = J', I(v^{-1}) \cap K = K', uv = w\}$$

$$= \sum_{K'' \subseteq K} \sum_{\substack{w \in W \\ I(w^{-1}) \cap K = K''}} w \cdot c(J, J', K, K', K'')$$

Where

$$c(J, J', K, K', K'') = \#\{(u,v) \in W^2 : D(u) \cap J = J', I(v^{-1}) \cap K = K', uv = w\}$$

is a constant whose existence is guaranteed by the previous theorem. Hence we have

$$X_{J'}^J X_{K'}^K = \sum_{K'' \subseteq K} c(J, J', K, K', K'') \sum_{\substack{w \in W \\ I(w^{-1}) \cap K = K''}} w$$

$$= \sum_{K'' \subseteq K} c(J, J', K, K', K'') X_{K''}^K$$

which is in the **Z**-span of $\#\{X_{K''}^K\}_{K'' \subseteq K}$, as we wanted.∎

Remark: Moszkowski and Solomon actually do more. They give interpretations for $c(J, J', K, K', K'')$ as cardinalities related to certain subgroups of W.

For our second application, we need to translate some of the combinatorics of words into the language of Coxeter groups.

Definition: Given a permutation $\sigma = \left(\begin{smallmatrix} 1 & \cdots & n \\ \sigma_1 & & \sigma_n \end{smallmatrix} \right) \in S_n$, we will say a word ω on letters $\{1, 2, \ldots, n\}$ is a *subword* of σ if $\omega = \sigma_{i_1} \cdots \sigma_{i_k}$ for some $1 \leq i_1 < \ldots < i_k \leq n$. We will say ω has *signature* $\mathcal{D}(\omega) = \{i : \omega_i > \omega_{i+1}\}$. For example, $\sigma = \left(\begin{smallmatrix} 123456 \\ 341265 \end{smallmatrix} \right)$ contains the subword $\omega = 416$, and we have $\mathcal{D}(\omega) = \{1\}$. Note that when we think of the permutation σ as a word, specifying its signature $\mathcal{D}(\sigma)$ is the same as specifying its descent set $D(\sigma)$.

Our goal is to prove a generalization of the following theorem of Kreweras and Moszkowski:

Theorem 3.3.6 ([KM], Théorème 3) *Fix a word ω of length k using letters from $\{1, 2, \ldots, n\}$ at most once, and also fix $J \subseteq \{1, 2, \ldots, n-1\}$. Then among all permutations $\sigma \in S_n$ with $\mathcal{D}(\sigma) = J$, the number which contain ω as a subword depends only on $\mathcal{D}(\omega)$.*

Our first task is to generalize the notion of "subwords" from S_n to other Coxeter groups.

Definition: A subgroup $W' \subseteq W$ is a *reflection subgroup* if W is generated by the reflections it contains, i.e. $W' = \langle W' \cap T \rangle$. Reflection subgroups W' share many of the properties enjoyed by parabolic subgroups W_J (see the Appendix). Among them is the following (Appendix, Proposition A.0.9): any $w \in W$ can be factored uniquely as a product $w = uv$ with $u \in W'$ and $I(v) \cap W' = \emptyset$. In this case, we say $u = \pi_{W'}(w)$.

Example: For $W = S_n$, a reflection subgroup corresponds to some partition of $\{1, 2, \ldots, n\}$ into blocks, and consists of all permutations of elements *within the same block*. For example

$$W' = S_{\{1,4,6\}} \times S_{\{2,3\}} \times S_{\{5\}}$$

is a reflection subgroup of S_6 (where $S_{\{1,4,6\}}$ is the subgroup permuting $1, 4, 6$ while fixing $2, 3, 5$), but the cyclic subgroup $\langle \left({123 \atop 231} \right) \rangle$ is *not* a reflection subgroup. Given $\sigma \in S_n$ and W', we can factor it into $\sigma = \pi_{W'}(\sigma)v$ as follows: v is obtained from σ by rearranging the numbers in each block (W'-orbit) of the partition to be in increasing order in the word $\sigma_1 \ldots \sigma_n$, and $\pi_{W'}(\sigma)$ is obtained by making the numbers in each block appear in the same order as they do in $\sigma_1 \ldots \sigma_n$, but subject to the constraint that $\pi_{W'}(\sigma) \in W'$. For example if

$$W' = S_{\{1,4,6\}} \times S_{\{2,3\}} \times S_{\{5\}} \text{ and } \sigma = \left({123456 \atop 341265} \right),$$

then $\sigma = \pi_{W'}(\sigma)v$ where

$$\pi_{W'}(\sigma) = \left({123456 \atop 432156} \right) \text{ and } v = \left({123456 \atop 214365} \right).$$

Key point: when the partition of $\{1, 2, \ldots, n\}$ corresponding to W' has only one non-singleton block $\{i_1, \ldots, i_k\}$, then the map $\pi_{W'} : S_n \rightarrow W'$ can be thought of as mapping σ to its *subword* ω on letters $\{i_1, \ldots, i_k\}$. Thus we have a way of thinking of subwords in terms of Coxeter group notions.

Having generalized subwords, we need a notion of when two subwords have the same "signature".

Definition: If W' is a reflection subgroup of W', then W' is a Coxeter group in its own right. In fact, if we define $\Phi_{W'}^+ = \{\alpha \in \Phi^+ : r_\alpha \in W'\}$, then it is possible to choose $\Pi_{W'} \subseteq \Phi_{W'}^+$ and $S' = \{r_\alpha : \alpha \in \Pi_{W'}\}$ so that (W', S') is a Coxeter system (Appendix Proposition A.0.8). Then for $w \in W'$ we let $D'(w) = \Pi_{W'} \cap w^{-1}(-\Phi_{W'}^+)$.

Example: Let $W = S_n$ and W' a reflection subgroup corresponding to the partition of $\{1, 2, \ldots, n\}$ into blocks B_i. Choose $\Pi_{W'}$ as follows: for each block $B_i = \{i_1, \ldots, i_k\}$ with $i_1 < \ldots < i_k$, we include

$$\{e_{i_1} - e_{i_2}, e_{i_2} - e_{i_3}, \ldots, e_{i_{k-1}} - e_{i_k}\}$$

in $\Pi_{W'}$. A moment's thought shows that if there is only one non-singleton block B_i, and if we think of $\pi_{W'}(\sigma)$ as a subword ω of σ on the letters in this block, then $D'(\pi_{W'}(\sigma))$ exactly encodes the same information as the signature $\mathcal{D}(\omega)$. For example, let $\sigma = \left(\begin{smallmatrix} 123456 \\ 341265 \end{smallmatrix}\right)$ and $W' = S_{\{1,4,6\}}$. Then $\pi_{W'}(\sigma) = \left(\begin{smallmatrix} 123456 \\ 423156 \end{smallmatrix}\right)$ has $D'(\pi_{W'}(\sigma)) = \{e_1 - e_4\}$, while $\omega = 416$ has $\mathcal{D}(\omega) = \{1\}$.

We can now prove our generalization of Theorem 3.3.6

Theorem 3.3.7 *Let W', W'' be two reflection subgroups of W, with $w\Pi_{W'} = \Pi_{W''}$ for some $w \in W$, and fix $J \subseteq S$. If $w' \in W'', w'' \in W''$ satisfy*

$wD'(w') = D''(w'')$ *(i.e. w' and w'' "have the same signature"), then*

$$\#\{u \in W : D(u) = J, \pi_{W'}(u) = w'\} = \#\{u \in W : D(u) = J, \pi_{W'}(u) = w'\}.$$

Proof: Define two parsets $P_{w'}, P_{w''}$ by $\overline{w'(\Pi_{W'}^+)}^{PLC}, \overline{w''(\Phi_{W''}^+)}^{PLC}$ respectively. Notice that

$$\begin{aligned}
\beta_J(\Sigma_{P_{w'}}) &= \#\{u \in \mathcal{L}(P_{w'}) : D(u) = J\} \\
&= \#\{u \in W : P_{w'} \subseteq u\Phi^+, D(u) = J\} \\
&= \#\{u \in W : w'(\overline{\Phi_{W'}^+}^{PLC}) \subseteq u\Phi^+, D(u) = J\} \\
&= \#\{u \in W : u^{-1}w'(\overline{\Phi_{W'}^+}^{PLC}) \subseteq \Phi^+, D(u) = J\} \\
&= \#\{u \in W : I(u^{-1}w') \cap W' = \emptyset \subseteq \Phi^+, D(u) = J\} \\
&= \#\{u \in W : D(u) = J, \pi_{W'}(u) = w'\}
\end{aligned}$$

and similarly for w'', so we need to show that $\beta_J(\Sigma_{P_{w'}}) = \beta_J(\Sigma_{P_{w''}})$. As in the previous application, we can apply Proposition 3.3.3, once we note that $w''ww'^{-1} \cdot w'\Pi_{W'} = w''\Pi_{W''}$ and

$$\begin{aligned}
w''ww'^{-1}(w'\Pi_{W'} \cap -\Phi^+) &= w''w(\Pi_{W'} \cap w'^{-1}(-\Phi^+)) \\
&= w''wD'(w') \\
&= w''D''(w'') \\
&= w''\Pi_{W''} \cap -\Phi^+
\end{aligned}$$

where the second equality comes from our hypotheses.∎

Corollary 3.3.8 *Theorem 3.3.6 holds.*

Proof: Given two words ω', ω'' with $\mathcal{D}(\omega') = \mathcal{D}(\omega'')$ and $J \subseteq S$ fixed, we want to show that among all permutations having descent set J, the number having ω' as a subword is the same as the number having ω'' as a subword. We want to apply the previous theorem with $W = S_n$, and W', W'' equal to the subgroups which permute the letters occurring in ω', ω'' respectively. If these sets of letters are $L' = \{i'_1, \ldots, i'_k\}, L'' = \{i''_1, \ldots, i''_k\}$ with $i'_1 < \ldots i'_k$ and $i''_1 < \ldots i''_k$, then we choose

$$\Pi_{W'} = \{e_{i'_1} - e_{i'_2}, e_{i'_2} - e_{i'_3}, \ldots, e_{i'_{k-1}} - e_{i'_k}\}$$

and similarly for $\Pi_{W''}$. We then choose w to be any permutation that takes i'_j to i''_j for all j. This means that $w\Pi_{W'} = \Pi_{W''}$, so we can apply the previous theorem. By the discussion in the preceding example, this gives the result.∎

Example: Let $W = S_6$, $\omega' = 416$, and $\omega'' = 425$. Then in the above proof, we choose $W' = S_{\{1,4,6\}}, W'' = S_{\{2,4,5\}}$,

$$\Pi_{W'} = \{e_1 - e_4, e_4 - e_6\}, \Pi_{W''} = \{e_2 - e_4, e_4 - e_5\},$$

and w is any permutation of the form $\left(\begin{smallmatrix} 123456 \\ 2**4*5 \end{smallmatrix}\right)$.

3.4 Multipartite P-partitions

In this section, we carry out a generalization of the theory of multipartite P-partitions ([GG], [Ge2]) to other Coxeter groups than S_n (as suggested in [Ge2], p. 300). We will need these results in Chapters 4 and 5 when we discuss quotients by diagonal embeddings of subgroups of W into W^r.

Let (W, S) be a finite Coxeter system realized by the positive root system (Φ, Π) in the vector space V. Fix $r \in \mathbf{P}$, and we now consider W acting on $V^r = \underbrace{V \times \cdots \times V}_{r \text{ times}}$ via

$$w(f_1, \ldots, f_r) = (w(f_1), \ldots, w(f_r)).$$

Definition: Order $\mathbf{R}^r$ *lexicographically*, i.e. let $(x_1, \ldots, x_r) \leq_{\mathcal{L}} (y_1, \ldots, y_r)$ if $\exists k \leq r - 1$ such that

$$x_1 = y_1, x_2 = y_2, \ldots, x_k = y_k \text{ and } x_{k+1} < y_{k+1}.$$

Given $(f_1, \ldots, f_r) \in V^r$, and $\alpha \in \Phi$, we will say $\langle \alpha, f \rangle \geq_{\mathcal{L}} \underline{0}$ if

$$(\langle \alpha, f_1 \rangle, \ldots, \langle \alpha, f_r \rangle) \geq_{\mathcal{L}} \underline{0}$$

(and similarly for $\langle \alpha, f \rangle >_{\mathcal{L}} \underline{0}$, etc.). For a parset P, we say $f \in V^r$ is an *r-partite P-partition* if

$$\langle \alpha, f \rangle \geq_{\mathcal{L}} \underline{0} \ \forall \alpha \in P$$

and

$$\langle \alpha, f \rangle >_{\mathcal{L}} \underline{0} \ \forall \alpha \in P \cap -\Phi^+$$

We denote the set of all r-partite P-partitions by $\mathcal{A}_r(P)$.

Example: Let $W = S_n$, $P = \Phi$. Then an r-partite P-partition $f = (f_1, \ldots, f_r)$ corresponds to a sequence of n vectors in $\mathbf{R}^r$ ordered lexicographically from largest to smallest. For example, let $n = 6$ and $r = 2$, and then $f = ((5, 4, 4, 3, 3, 2), (1, 3, 2, 3, 3, 3))$ corresponds to

$$\binom{5}{1} \geq_{\mathcal{L}} \binom{4}{3} \geq_{\mathcal{L}} \binom{4}{2} \geq_{\mathcal{L}} \binom{3}{3} \geq_{\mathcal{L}} \binom{3}{3} \geq_{\mathcal{L}} \binom{2}{3}.$$

If $n =, r = 2$ and P is the parset from Figure 2, then

$$f = ((f_{11}, f_{12}, f_{13}), (f_{21}, f_{22}, f_{23}))$$

is in $\mathcal{A}_2(P)$ when

$$\binom{f_{12}}{f_{22}} >_{\mathcal{L}} \binom{f_{11}}{f_{21}} \quad \text{and} \quad \binom{f_{12}}{f_{22}} \geq_{\mathcal{L}} \binom{f_{13}}{f_{23}}$$

Proposition 3.4.1

$$\mathcal{A}_r(P) = \coprod_{w \in \mathcal{L}(P)} \mathcal{A}_r(w\Phi^+)$$

Proof: Same as Proposition 3.1.1 (which is the $r = 1$ case). The only properties we used there were:

1. The linear maps $\langle \alpha, \cdot \rangle : V \to \mathbf{R}$ are well-defined $\forall \alpha \in \Phi$.

2. $\mathbf{R}$ is a totally ordered vector space.

Replacing V by V^r and $\mathbf{R}$ by $\mathbf{R}^r$, these properties still hold, and the proof goes through.∎

Theorem 3.4.2 (cf. [Ge2], Theorem 16)

$$\mathcal{A}_r(w\Phi^+) =$$

$$\coprod_{\substack{(w_1, \ldots, w_r) \in W^r \\ w_r w_{r-1} \cdots w_1 = w}} \#\{(f_1, \ldots, f_r) \in V^r : (w_r w_{r-1} \cdots w_{i+1})^{-1}(f_i) \in \mathcal{A}(w_i \Phi^+) \ \forall i\}$$

To prove this theorem, we mimic the proof of Theorem 16 in [Ge2], and first prove a lemma which is slightly more general than the case $r = 2$:

Lemma 3.4.3 (cf. [Ge2], Theorem 9) *Let V_1, V_2 be two vector spaces with W-actions, and R_1, R_2 two totally ordered vector spaces, with linear maps $\langle \alpha, \cdot \rangle : V_i \to R_i$ for $i = 1, 2$. Put the lexicographic total order on $R_1 \times R_2$ (as in the previous definition), and then define $\mathcal{A}_{V_1}(P), \mathcal{A}_{V_2}(P), \mathcal{A}_{V_1 \times V_2}(P)$ as before. Then*

$$\mathcal{A}_{V_1 \times V_2}(w\Phi^+) =$$

$$\coprod_{\substack{(w_1, w_2) \\ w_2 w_1 = w}} \{(f_1, f_2) \in V_1 \times V_2 : f_2 \in \mathcal{A}_{V_2}(w_2 \Phi^+), w_2^{-1}(f_1) \in \mathcal{A}_{V_1}(w_1 \Phi^+)\}$$

Proof: First we check that the sets on the right are actually disjoint. So suppose $(f_1, f_2) \in V_1 \times V_2$ satisfies

$$f_2 \in \mathcal{A}_{V_2}(w_2 \Phi^+) \cap \mathcal{A}_{V_2}(v_2 \Phi^+)$$

and

$$w_2^{-1}(f_1) \in \mathcal{A}_{V_1}(w_1 \Phi^+) \cap \mathcal{A}_{V_1}(v_1 \Phi^+)$$

for some $w_1, w_2, v_1, v_2 \in W$. By the analogue of Proposition 3.1.1 for V_2, the first line allows us to conclude that $w_2 = v_2$. But then $w_2^{-1}(f_1) = v_2^{-1}(f_1)$, so the second line allows us to conclude that $w_1 = v_1$. Thus the sets on the right are disjoint.

Furthermore, the sets on the right cover all of $V_1 \times V_2$, as w ranges over all of W. To see this, let $(f_1, f_2) \in V_1 \times V_2$. By the analogue of Proposition 3.1.1 for V_2, we know $f_2 \in \mathcal{A}_{V_2}(w_2 \Phi^+)$ for some $w_2 \in W$, and then we know $w_2^{-1}(f_1) \in \mathcal{A}_{V_1}(w_1 \Phi^+)$ for some $w_1 \in W$. Hence (f_1, f_2) lies in the set on the right corresponding to (w_1, w_2).

Thus it only remains to show that each of the sets on the right is actually in $\mathcal{A}_{V_1 \times V_2}(w\Phi^+)$, i.e. we need to show that

$$f_2 \in \mathcal{A}_{V_2}(w_2 \Phi^+), w_2^{-1}(f_1) \in \mathcal{A}_{V_1}(w_1 \Phi^+), \text{ and } w_2 w_1 = w$$

imply

1. $\langle \alpha, f_1 \rangle \geq 0 \ \forall \alpha \in w\Phi^+$

2. $\langle \alpha, f_1 \rangle = 0$ for some $\alpha \in w\Phi^+ \Rightarrow$

   $\langle \alpha, f_2 \rangle \geq 0$ and

   $\langle \alpha, f_2 \rangle > 0$ if $\alpha \in w\Phi^+ \cap -\Phi^+$

To prove 1, note that

$$w_2^{-1}(f_1) \in \mathcal{A}_{V_1}(w_1\Phi^+) \ \Rightarrow \ \langle \beta, w_2^{-1}(f_1) \rangle \geq 0 \ \forall \beta \in w_1(\Phi^+),$$
$$\langle \beta, w_2^{-1}(f_1) \rangle > 0 \ \forall \beta \in w_1(\Phi^+) \cap -\Phi^+$$

$$\Rightarrow \ \langle w_2(\beta), f_1 \rangle \geq 0 \ \forall w_2(\beta) \in w_2 w_1(\Phi^+),$$
$$\langle w_2(\beta), f_1 \rangle > 0 \ \forall w_2(\beta) \in w_2 w_1(\Phi^+) \cap w_2(-\Phi^+)$$

$$\Rightarrow \ \langle \alpha, f_1 \rangle \geq 0 \ \forall \alpha \in w\Phi^+,$$
$$\langle \alpha, f_1 \rangle > 0 \ \forall \alpha \in w\Phi^+ \cap w_2(-\Phi^+).$$

This proves assertion 1 (and a bit more). To prove the first assertion in 2, assume $\langle \alpha, f_1 \rangle = 0$ for some $\alpha \in w\Phi^+$. Then $\alpha \notin w_2(-\Phi^+)$, so $\alpha \in w_2(\Phi^+)$, and hence $\langle \alpha, f_2 \rangle \geq 0$ since $f_2 \in \mathcal{A}_{V_2}(w_2\Phi^+)$.

To prove the second assertion in 2, assume that this same α is in $w\Phi^+ \cap -\Phi^+$. Then $w_2^{-1}(\alpha) \in w_2^{-1}(-\Phi^+)$. Since $w_2^{-1}(\alpha) \in \Phi^+$ (see the previous paragraph), we can write $w_2^{-1}(\alpha) = \sum c_i \alpha_i$ with $c_i > 0$ and $\alpha_i \in \Pi$. Hence we must have $\alpha_i \in w_2^{-1}(-\Phi^+)$ for some i_0 (else $w_2^{-1}(\alpha) \notin w_2^{-1}(-\Phi^+)$). Therefore

$$\langle \alpha, f_2 \rangle = c_{i_0} \langle w_2(\alpha_{i_0}), f_2 \rangle + \sum_{i \neq i_0} c_i \langle w_2(\alpha_i), f_2 \rangle > 0$$

since $f_2 \in \mathcal{A}_{V_2}(w_2\Phi^+)$. ■

Proof of Theorem 3.4.2: We use induction on r. The case $r = 1$ is trivial, and $r = 2$ is the specialization of the previous lemma to

$$V_1 = V_2 = V, R_1 = R_2 = \mathbf{R}.$$

For $r \geq 3$, we have

$\mathcal{A}_r(w\Phi^+)$

$$= \mathcal{A}_{V \times V^{r-1}}(w\Phi^+)$$

$$= \coprod_{\substack{(w_1,w_2)\in W^2 \\ w_2 w_1 = w}} \left\{ \begin{array}{c} (f_1,(f_2,...,f_r))\in V \times V^{r-1}: \\ (f_2,...,f_r)\in\mathcal{A}_{r-1}(w_2\Phi^+), w_2^{-1}(f_1)\in\mathcal{A}(w_1\Phi^+) \end{array} \right\}$$

$$= \coprod_{\substack{(w_1,w_2)\in W^2 \\ w_2 w_1 = w}} \coprod_{\substack{(w'_2,w'_3,...,w'_r)\in W^{r-1} \\ w'_r w'_{r-1}\cdots w'_2 = w_2}} \left\{ \begin{array}{c} (f_1,f_2,...,f_r)\in V^r:\forall i\geq 2 \\ (w'_r w'_{r-1}\cdots w'_{i+1})^{-1}(f_i)\in\mathcal{A}(w_i\Phi^+), w_2^{-1}(f_1)\in\mathcal{A}(w_1\Phi^+) \end{array} \right\}$$

$$= \coprod_{\substack{(w_1,w_2,...,w_r)\in W^r \\ w_r w_{r-1}\cdots w_1 = w}} \left\{ \begin{array}{c} (f_1,f_2,...,f_r)\in V^r:\forall i \\ (w_r w_{r-1}\cdots w_{i+1})^{-1}(f_i)\in\mathcal{A}(w_i\Phi^+) \end{array} \right\}.$$

The second equality above comes from the previous lemma applied with $V_1 = V, V_2 = V^{r-1}, R_1 = \mathbf{R}, R_2 = \mathbf{R}^{r-1}$. The third equality is by the induction hypothesis. The fourth equality is because $w_2^{-1} = (w'_r w'_{r-1} \cdots w'_2)^{-1}$. ■

Just as we used Proposition 3.1.1 to partition and shell $\Sigma(W,S)$ in Section 3.2, we will now use Theorems 3.4.1, 3.4.2 to partition and shell $\Sigma(W^r, rS)$. We now consider W^r acting on V^r by

$$(w_1,\ldots,w_r)(f_1,\ldots,f_r) = (w_1(f_1),\ldots,w_r(f_r)).$$

Given $f = (f_1,\ldots,f_r)$, we can extend the definition of our map F (from Section 3.1) by setting

$$F(f) = (w_1,\ldots,w_r)W_{(J_1,\ldots,J_r)} = w_1 W_{J_1} \times \cdots \times w_r W_{J_r}$$

where $w_i W_{J_i} = F(f_i)$.

Theorem 3.4.4

$$\Sigma(W^r, rS) = \coprod_{(w_1,\ldots,w_r)\in W^r} \prod_{i=1}^{r}[w_r w_{r-1}\cdots w_i W_{S-D(w_i)}, w_r w_{r-1}\cdots w_i W_{\emptyset}]$$

is a partitioning of $\Sigma(W^r, rS)$.

Proof: Applying Propositions 3.4.1, 3.4.2 to the empty parset $P = \emptyset$ gives

$$
\begin{aligned}
V^r &= \coprod_{w\in W} \mathcal{A}_r(w\Phi^+) \\
&= \coprod_{(w_1,\ldots,w_r)\in W^r} \{(f_1, f_2,\ldots, f_r) \in V^r : \forall i \\
&\qquad (w_r w_{r-1}\cdots w_{i+1})^{-1}(f_i) \in \mathcal{A}(w_i\Phi^+) \}
\end{aligned}
$$

Applying the operation $F \mapsto F(f)$ to both ends of the above equation gives

$$\bigcup_{f\in V^r} F(f) = \bigcup_{(w_1,\ldots,w_r)\in W^r} \{F((f_1,\ldots,f_r)) : \forall i$$
$$(w_r w_{r-1}\cdots w_{i+1})^{-1}(f_i) \in \mathcal{A}(w_i\Phi^+) \}$$

which in light of Lemma 3.2.1 gives

$$\Sigma(W^r, rS) = \coprod_{(w_1,\ldots,w_r)\in W^r} \prod_{i=1}^{r}[w_r w_{r-1}\cdots w_i W_{S-D(w_i)}, w_r w_{r-1}\cdots w_i W_{\emptyset}]$$

as desired.■

We now put a shelling order on the facets of $\Sigma(W^r, rS)$.

Definition: We say $(w_1,\ldots, w_r) <_{\mathcal{RLW}} (w_1',\ldots, w_r')$ in *reverse lexicographic weak order*, if there exists $k \geq 2$ such that

$$w_r = w_r', w_{r-1} = w_{r-1}',\ldots, w_k = w_k' \text{ and } w_{k-1} <_R w_{k-1}'$$

Theorem 3.4.5 *Order* W^r *by any linear extension of* $<_{\mathcal{RLW}}$. *Then the partitioning in the previous theorem is a shelling.*

Proof: It suffices to show that if we have $(w_1, \ldots, w_r), (w'_1, \ldots, w'_r)$ satisfying

$$w'_r w'_{r-1} \cdots w'_i W_\emptyset \in w_r w_{r-1} \cdots w_i W_{S-D(w_i)} \ \forall i$$

then $(w_1, \ldots, w_r) <_{\mathcal{RLW}} (w'_1, \ldots, w'_r)$.

Since $w'_r \in w_r W_{S-D(w_r)}$, we have $\pi_{S-D(w_r)}(w'_r) = w_r$ and hence $w_r \leq_R w'_r$. If $w_r <_R w'_r$, then we're done, so assume $w_r = w'_r$. Then from $w'_r w'_{r-1} W'_\emptyset \in w_r w_{r-1} W_{S-D(w_{r-1})}$, we conclude that $w'_{r-1} \in w_{r-1} W_{S-D(w_{r-1})}$ and hence $w_{r-1} \leq_R w'_{r-1}$. Continuing this process, we eventually get $(w_1, \ldots, w_r) <_{\mathcal{RLW}} (w'_1, \ldots, w'_r)$.∎

Remark: There is a much more straightforward partitioning and shelling based on the fact the $\Sigma(W^r, rS) = \Sigma(W, S) * \cdots * \Sigma(W, S)$, and $\Sigma(W, S)$ is shellable. However, this partitioning will not be as useful for our purposes, because it does not behave as nicely with respect to the diagonal action of W on $\Sigma(W^r, rS)$.

Example: Let $(W, S) = (S_2, \{(12)\})$ in its usual realization as permuting coordinates on $\mathbf{R}^2$, and let $r = 2$. What do some of the theorems of this section say in this case? Choosing $P = \emptyset$, Proposition 3.4.1 says that

$$\{((f_{11}, f_{12}), (f_{21}, f_{22})) \in \mathbf{R}^2 \times \mathbf{R}^2\} = \left\{ \begin{pmatrix} f_{11} \\ f_{21} \end{pmatrix} \geq_{\mathcal{L}} \begin{pmatrix} f_{12} \\ f_{22} \end{pmatrix} \right\} \bigsqcup \left\{ \begin{pmatrix} f_{12} \\ f_{22} \end{pmatrix} >_{\mathcal{L}} \begin{pmatrix} f_{11} \\ f_{21} \end{pmatrix} \right\}$$

and Theorem 3.4.2 refines this further as

$$\left\{ \begin{pmatrix} f_{11} \\ f_{21} \end{pmatrix} \geq_{\mathcal{L}} \begin{pmatrix} f_{12} \\ f_{22} \end{pmatrix} \right\} = \{f_{11} \geq f_{12}, f_{21} \geq f_{22}\} \amalg \{f_{11} > f_{12}, f_{21} < f_{22}\}$$

$$\left\{ \begin{pmatrix} f_{12} \\ f_{22} \end{pmatrix} >_{\mathcal{L}} \begin{pmatrix} f_{11} \\ f_{21} \end{pmatrix} \right\} = \{f_{12} > f_{11}, f_{21} \geq f_{22}\} \amalg \{f_{12} \geq f_{11}, f_{22} > f_{21}\}$$

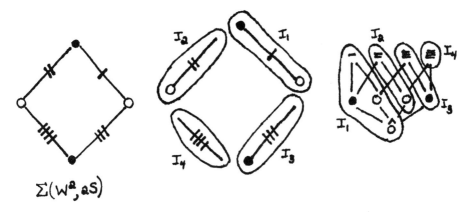

$\Sigma(W^2, 2S)$

Figure 3.4: Shelling of $\Sigma(W^2, 2S)$ for $(W, S) = (S_2, \{(12)\})$

The shelling and partitioning asserted by the last two theorems goes as follows:

$$
\begin{aligned}
\Sigma(W^2, 2S) \;=\;& [(id, id)W_{(S,S)}, (id, id)W_{(\emptyset,\emptyset)}] \\
\sqcup\; & [((12), id)W_{(\emptyset,S)}, ((12), id)W_{(\emptyset,\emptyset)}] \\
\sqcup\; & [((12), (12))W_{(S,\emptyset)}, ((12), (12))W_{(\emptyset,\emptyset)}] \\
\sqcup\; & [(id, (12))W_{(\emptyset,\emptyset)}, (id, (12))W_{(\emptyset,\emptyset)}]
\end{aligned}
$$

Figure 4 shows how this decomposes $\Sigma(W^2, 2S)$ as a simplicial poset and its topological realization (where we have labelled the intervals in this shelling in order as I_1, I_2, I_3, I_4).

Chapter 4

Quotients by reflection and alternating subgroups, and their diagonal embeddings

4.1 Reflection subgroups and their diagonal embeddings

In this chapter, we return to the subject of quotients $\Sigma(W, S)/G$ and study some specific classes of subgroups G. This section deals with *reflection subgroups* along with their *diagonal embeddings* in W^r. Recall that $W' \subseteq W$ is a reflection subgroup if it is generated by the reflections it contains.

Definition: The *diagonal embedding* $\Delta^r : W \to W^r$ is the map given by $\Delta^r(w) = (w, \ldots, w)$. Given a subgroup $G \subseteq W$, let $\Delta^r(W')$ denote the subgroup of W^r which is the image of G under Δ^r. It turns out that the theory of P-paritions and r-partite P-partitions developed in Chapter 3 will help us to understand the quotients $\Sigma(W^r, rS)/\Delta^r(W')$ by providing us with a fundamental domain for the action of $\Delta^r(W')$ on V^r.

Definition: Given a reflection subgroup $W' \subseteq W$, define the parset $P(W') = \overline{\Phi_{W'}^+}^{PLC}$ (recall that $\Phi_{W'}^+ = \{\alpha \in \Phi^+ : r_\alpha \in W'\}$). It is clear from the definitions that

$$\mathcal{L}(P(W')) = \{w \in W : I(w) \cap W' = \emptyset\}.$$

Proposition 4.1.1 $\mathcal{A}_r(P(W'))$ *is a fundamental domain* V^r *for the action of* $\Delta^r(W')$, *i.e. every orbit* $W'f$ *of a vector* $f \in V^r$ *has a unique representative in* $\mathcal{A}_r(P(W'))$.

For the proof, we require a lemma giving a multipartite generalization of the fact after the first definition in Section 3.2.

Lemma 4.1.2 *Let the* multipartite fundamental chamber $\mathcal{C}_r$ *be defined by*

$$\mathcal{C}_r = \{f \in V^r : \langle \alpha, f \rangle \geq_{\mathcal{L}} \underline{0} \ \forall \alpha \in \Phi^+\}$$

Then $w(f) \in \mathcal{C}_r$ *implies that*

1. $\{v \in W : v(f) \in \mathcal{C}_r\} = W_J w$ *where*

$$J = \{\alpha \in \Pi : \langle \alpha, w(f) \rangle = \underline{0}\}$$

(Notice that we are abusing notation in our usual way by not distinguishing between J *and* $\{r_\alpha : \alpha \in J\}$).

    2. $v(f) \in \mathcal{C}_r$ if only if $v(f) = w(f)$ (i.e. the W-translate of f lying in $\mathcal{C}_r$ is unique).

Proof: Note that $\alpha \in J \Rightarrow r_\alpha w(f) = w(f)$, since r_α fixes all vectors orthogonal to α. This shows that $W_J w \subseteq \{v \in W : v(f) \in \mathcal{C}_r\}$, and also that 1 implies 2.

Thus we need only show that the reverse inclusion holds in 1. Let $f = (f_1, \ldots, f_r)$ and suppose $v(f) \in \mathcal{C}_r$. Looking at first coordinates, this implies $v(f_1), w(f_1) \in \mathcal{C}$. Then by the standard (non-multipartite, $r = 1$) version of this lemma, we conclude that $v(f_1) = w(f_1)$ and $v = uw$ for some $u \in W_{K_1}$ where

$$K_1 = \{\alpha \in \Pi : \langle \alpha, w(f_1) \rangle = 0\}.$$

Now let V_{K_1} be the linear span of K_1, and let $\pi_{K_1} : V \to V_{K_1}$ be orthogonal projection onto V_{K_1} (with respect to $\langle \cdot, \cdot \rangle$). Note that (W_{K_1}, K_1) forms a Coxeter system on V_{K_1} with simple roots K_1. For all $\alpha \in K_1$, we have

$$\langle \alpha, \pi_{K_1}(w(f_2)) \rangle = \langle \alpha, w(f_2) \rangle \geq 0$$

since $\alpha \in \Phi^+$, and similarly for $v(f_2)$. So by applying the standard version of this lemma again (this time to the Coxeter system (W_{K_1}, K_1)), we conclude that

$$
\begin{aligned}
\pi_{K_1}(w(f_2)) &= \pi_{K_1}(v(f_2)) \\
&= \pi_{K_1}(uw(f_2)) \\
&= u\pi_{K_1}(w(f_2))
\end{aligned}
$$

The last equality holds because $u \in W_{K_1}$ implies u commutes with π_{K_1}.

Applying the standard version of this lemma again tells us that $u \in W_{K_2}$ where

$$K_2 = \{\alpha \in K_1 : \langle \alpha, \pi_{K_1}(w(f_2)) \rangle = 0\}$$

$$= \{\alpha \in K_1 : \langle \alpha, w(f_2) \rangle = 0\}$$

Repeating this process, we eventually conclude that $u \in W_K$, where

$$
\begin{aligned}
K' &= \{\alpha \in \Pi : \langle \alpha, w(f_1) \rangle = \cdots \langle \alpha, w(f_r) \rangle = 0\} \\
&= \{\alpha \in \Pi : \langle \alpha, w(f) \rangle = \underline{0}\} \\
&= J
\end{aligned}
$$

This shows that $v \in W_J u$, as we wanted. ∎

Proof of Proposition 4.1.1: ¿From Proposition 3.4.1, we know that

$$\mathcal{A}_r(P(W')) = \coprod_{w \in \mathcal{L}(P(W'))} \mathcal{A}_r(w\Phi^+)$$

and we also know that

$$\mathcal{L}(P(W')) = \{w \in W : I(w) \cap W' = \emptyset\}.$$

Thus we need to show that in each orbit $W'f$ there exists a unique e such that $e \in \mathcal{A}_r(w\Phi^+)$ for some w with $I(w) \cap W' = \emptyset$.

Existence: Given $e \in W'f$, we know that $e \in \mathcal{A}_r(w\Phi^+)$ for some $w \in W$ (by 3.4.1 applied to $P = \emptyset$). So choose $e \in W'f$ such that $l(w)$ *is minimal*, and we will show by contradiction that $I(w) \cap W' = \emptyset$.

Assume not, i.e. let $r_\beta \in I(w) \cap W$ for some $\beta \in \Phi^+$. Let v satisfy $r_\beta(e) \in \mathcal{A}_r(v\Phi^+)$ (we know such a v exists). Our strategy will be to show that $l(v) < l(w)$, and hence get a contradiction. We have $v^{-1}r_\beta(e) \in \mathcal{C}_r$, so by the previous lemma, we have that $v^{-1}r_\beta(e) = w^{-1}(e)$ and $r_\beta v = wu$ for some $u \in W_K$ where

$$K = \{\alpha \in \Pi : \langle \alpha, v^{-1}r_\beta(e) \rangle = \underline{0}\}.$$

Furthermore, since $\langle \alpha, v^{-1}r_\beta(e) \rangle >_{\mathcal{L}} \underline{0} \; \forall \alpha \in D(v)$, we must have $K \subseteq S - D(v)$. Thus $l(v) \le l(vu^{-1})$, since $u^{-1} \in W_K \subseteq W_{S-D(v)}$ implies $v = \pi_K(vu^{-1})$. But $vu^{-1} = r_\alpha w$, and $l(r_\alpha w) < l(w)$ since $r_\alpha \in I(w)$. Hence $l(v) < l(w)$, as we wanted.

Uniqueness: Suppose $e \in \mathcal{A}_r(w_1\Phi^+)$ and $w' \in \mathcal{A}_r(w_2\Phi^+)$ for some $w' \in W$ and $I(w_i) \cap W' = \emptyset$ for $i = 1, 2$. Then $w_2^{-1}w'(e), w_1^{-1}(e) \in \mathcal{C}_r$, so by the previous lemma we conclude that $w_2^{-1}w'(e) = w_1^{-1}(e)$ and $w'^{-1}w_2 = w_1u$ for some $u \in W_K$ where $K \subseteq D(w_1) \cap D(w_2)$. Thus $w_2 = w'w_1u$, and hence we have $w_2 = w_1$, since in each double coset $W'wW_K$ there is a unique element w satisfying $I(w) \cap W' = \emptyset$ and $K \subseteq S - D(w)$ (by Proposition A.0.11). Thus $w'(e) = e.\blacksquare$

Example: The previous theorem is nearly trivial for $W = S_n$. Recall that in this case a reflection subgroup W' is the set of all permutations within the blocks of some partition π of $\{1, 2, \ldots, n\}$. It is easy to see that a vector $f \in \mathcal{A}_r(P(W'))$ corresponds to a sequence of n vectors $(v_1, \ldots, v_n)$ in $\mathbf{R}^r$ in which the v_i's corresponding to the same block of π appearing in decreasing order lexicographically. Thus our theorem states the obvious fact that we can use W' to permute $(v_1, \ldots, v_n)$ uniquely so as to make this condition hold. For example, if $r = 2, n = 6$ and $W' = S_{\{1,4,5\}} \times S_{\{2,6\}} \times S_{\{3\}}$, then for every $f = ((f_{11}, \ldots, f_{16}), (f_{21}, \ldots, f_{26}))$ there is a unique element $e \in W'f$

satisfying

$$\begin{pmatrix} e_{11} \\ e_{21} \end{pmatrix} \geq_{\mathcal{L}} \begin{pmatrix} e_{14} \\ e_{24} \end{pmatrix} \geq_{\mathcal{L}} \begin{pmatrix} e_{15} \\ e_{25} \end{pmatrix} \text{ and }$$

$$\begin{pmatrix} e_{12} \\ e_{22} \end{pmatrix} \geq_{\mathcal{L}} \begin{pmatrix} e_{16} \\ e_{26} \end{pmatrix}.$$

Theorem 4.1.3

$$\Sigma(W^r, rS)/\dot{\Delta}^r(W') =$$

$$\coprod_{\substack{(w_1,\ldots,w_r) \in W^r \\ I(w_r w_{r-1} \cdots w_1) \cap W' = \emptyset}} \Delta^r(W') \prod_{i=1}^r [w_r w_{r-1} \cdots w_i W_{S-D(w_i)}, w_r w_{r-1} \cdots w_i W_\emptyset]$$

is a partitioning.

Proof: Let V^r/W' denote the set of orbits $W'f$ of all vectors $f \in V^r$. Then from the previous theorem we conclude that

$$V^r/W' = \{W'f : f \in \mathcal{A}_r(P(W'))\}$$

$$= \coprod_{\substack{(w_1,\ldots,w_r) \in W^r \\ I(w_r w_{r-1} \cdots w_1) \cap W' = \emptyset}} \{W'(f_1,\ldots,f_r) : w_r w_{r-1} \cdots w_{i+1}(f_i) \in \mathcal{A}(w_i \Phi^+)\}$$

where the second equality follows from Propositions 3.4.1,3.4.2. For any orbit Gf, define $F(Gf) = \bigcup_{g(f) \in Gf} F(g(f))$ and note that $F(Gf) = GF(f)$. Thus applying the map $W'f \mapsto F(W'f)$ to both ends of the above equation (and using Lemma 3.2.1) gives

$$\Sigma(W^r, rS)/\Delta^r(W') =$$

$$\coprod_{\substack{(w_1,\ldots,w_r) \in W^r \\ I(w_r w_{r-1} \cdots w_1) \cap W' = \emptyset}} \Delta^r(W') \prod_{i=1}^r [w_r w_{r-1} \cdots w_i W_{S-D(w_i)}, w_r w_{r-1} \cdots w_i W_\emptyset]$$

as we wanted. ∎

This theorem allows us to give a combinatorial interpretation to the β_J's of the quotient complex $\Sigma(W^r, rS)/\Delta^r(W')$:

Corollary 4.1.4

$$\beta_{J_1,\ldots,J_r} = \#\{(w_1,\ldots,w_r) : D(w_i) = J_i, I(w_r w_{r-1} \cdots w_1) \cap W' = \emptyset\}$$

Proof: see Proposition 3.2.3. ∎

Corollary 4.1.5 *If r is even then* $\forall J_1, \ldots, J_r \subseteq S$ *we have*

$$\#\{(w_1,\ldots,w_r) \in W^r : I(w_r \cdots w_1) \cap W' = \emptyset, D(w_i) = J_i\}$$

$$= \#\{(w_1,\ldots,w_r) \in W^r : I(w_r \cdots w_1) \cap W' = \emptyset, D(w_i) = S - J_i\}$$

Proof: When r is even, $\text{sgn}(w,\ldots,w) = \text{sgn}(w)^r = 1 \ \forall w \in W'$. Hence by Proposition 2.4.4, we have $\beta_{J_1,\ldots,J_r} = \beta_{S-J_1,\ldots,S-J_r} \ \forall J_1,\ldots,J_r \subseteq S$. Now apply the previous corollary. ∎

Remark: We do not know how to prove this last corollary bijectively. However, Gessel (personal communication) has shown how to prove an even stronger result for the special case of $W = S_n$ using the theory of *symmetric functions* and their *canonical involution*.

For shellability results, we require another partial order on W and W^r.

Definition: The *(strong) Bruhat order* $<_B$ on W is defined to be the transitive closure of the relations $wt <_B w$ if $w \in W, t \in T$ and $l(tw) < l(w)$.

We will say $(w_1, \ldots, w_r) <_{\mathcal{RLB}} (w_1', \ldots, w_r')$ in *reverse lexicographic Bruhat order* if for some $k \geq 2$ we have

$$w_r = w_r', w_{r-1} = w_{r-1}', \ldots, w_k = w_k' \text{ and } w_{k-1} <_B w_{k-1}'.$$

Theorem 4.1.6 *For $r = 1, 2$, if we order W^r by any linear extension of $<_{\mathcal{RLB}}$, then the partitioning of the previous theorem is a shelling.*

Proof: For $r = 1$, we need to show that if w_1, w_2 both satisfy $I(w_i) \cap W' = \emptyset$, then

$$W' w_1 W_\emptyset \subseteq W' w_2 W_{S-D(w_2)} \Rightarrow w_2 \leq_B w_1.$$

But this follows immediately from Proposition A.0.11 which says that the unique element $w \in W' w W_J$ satisfying $I(w) \cap W' = \emptyset, D(w) \subseteq S - J$ is the least element of $W' w W_J$ in Bruhat order.

For $r = 2$, we need to that if $(u_1, u_2), (v_1, v_2)$ satisfy

1. $I(u_2 u_1) \cap W' = I(v_2 v_1) \cap W' = \emptyset$

2. $\Delta^2(v_2, v_1) \subseteq \Delta^2(u_2, u_1) W_{(S-D(u_1), S-D(u_2))}$

then $(u_2, u_1) \leq_{\mathcal{RLB}} (v_1, v_2)$. We thank M. Dyer for supplying the proof of a slightly stronger technical lemma, which appears in the Appendix as Lemma A.0.14.■

Corollary 4.1.7 *For $r = 1, 2$, $\Sigma(W^r, rS)/\Delta^r(W')$ is CM/k for all fields k.*

Proof: see the remarks after the definition of shellability in Chapter 3.■

Remark: It is easy to see by example that the previous theorem is tight, in the sense that $\Sigma(W^r, rS)/\Delta^r(W')$ may be non-shellable for $r \geq 3$. In fact, we have already seen that if $(W, S) = (\mathbf{Z}_2, \{s\})$ and $W' = W$, then $\Sigma(W^r, rS)/\Delta^r(W')$ is homeomorphic to $\mathbf{R}P^{r-1}$. For $r \geq 3$, this is not Cohen-Macaulay over fields of characteristic 2, and hence non-shellable.

Theorem 4.1.8

1. $\Sigma(W, S)/W'$ *is homeomorphic to an* $(\#S - 1)$*-ball.*

2. $\Sigma(W^2, 2S)/\Delta^2(W')$ *is homeomorphic to a* $(2\#S - 1)$*-sphere.*

Proof: We use a fact which is a special case of ([Bj1], Proposition 4.3): If P is a shellable simplicial poset which is also a pseudomanifold with boundary, then P is homeomorphic is either to a sphere or disk, depending on whether P is a pseudomanifold or not. When $r = 1, 2$, from the previous theorem we know that $\Sigma(W^r, rS)/\Delta^r(W')$ is shellable, and from Proposition 2.4.2 we know that $\Sigma(W, S)/G$ is always a pseudomanifold with boundary, and a pseudomanifold if and only if $G \cap T = \emptyset$. It is easy to see that $W' \cap T \neq \emptyset$, while $\Delta^2(W') \cap T = \emptyset$, so the result follows.∎

Example: As noted earlier, for $(W, S) = (\mathbf{Z}_2, \{s\})$ and $W' = W$, then the quotient $\Sigma(W^r, rS)/\Delta^r(W')$ is homeomorphic to $\mathbf{R}P^{r-1}$. Notice that $\mathbf{R}P^0$ is a ball, and $\mathbf{R}P^1$ is a sphere, in agreement with our last theorem.

4.2   Application: invariants of permutation groups

We now return to the application mentioned in the introduction which motivated much of this work. We beg the reader's pardon in advance for the seemingly unavoidable use of multi-indices.

Definition: Let

$$\mathcal{R}_r = \underbrace{\mathbf{Q}[x_1,\ldots,x_n] \otimes \cdots \otimes \mathbf{Q}[x_1,\ldots,x_n]}_{r\text{-fold tensor product}}$$

$$\cong \mathbf{Q}[x_1^{(1)},\ldots,x_n^{(1)},x_1^{(2)},\ldots,x_n^{(2)},\cdots,x_1^{(r)},\ldots,x_n^{(r)}]$$

and let permutations $\sigma \in S_n$ act on $\mathcal{R}_r$ by permuting the variables $x_i^{(j)}$ as follows: $\sigma(x_i^{(j)}) = x_{\sigma(i)}^{(j)}$.

Given any subgroup G of S_n, G also acts on $\mathcal{R}_r$, and our problem is to find a certain "nice decomposition" of the invariant subring $\mathcal{R}_r^G$ (suggested by Gessel in the case $G = S_n$). In order to say what this "nice description" is, we need a few more definitions.

Definition: Define an $\mathbf{N}^r$-grading on $\mathcal{R}_r$ by setting

$$\deg(x_i^{(j)}) = \epsilon_j = \text{ the } j^{th} \text{ standard basis vector in } \mathbf{N}^r.$$

Note that our S_n-action preserves this grading. For an $\mathbf{N}^r$-graded $\mathbf{Q}$-algebra Q, let its *Hilbert series* $F(Q,t)$ be the formal power series in the variables $t^{(1)},\ldots,t^{(r)}$ given by

$$F(Q,t) = \sum_{\alpha \in \mathbf{N}^r} dim_{\mathbf{Q}} Q_\alpha \cdot t^\alpha,$$

where $t^\alpha = (t^{(1)})^{\alpha_1} \cdots (t^{(r)})^{\alpha_r}$ if $\alpha = (\alpha_1, \ldots, \alpha_r)$. One nice description of $\mathcal{R}_r^G$ that we seek is its Hilbert series $F(\mathcal{R}_r^G, t)$.

Definition: Let

$$e_i(x^{(j)}) = \sum_{I \subseteq \{1,\ldots,n\}, \#I=i} \prod_{l \in I} x_l^{(j)}$$

be the ith *elementary symmetric function* in the variables $x_1^{(i)}, \ldots, x_r^{(i)}$. It is easy to see that $e_i(x^{(j)}) \subseteq \mathcal{R}_r^{S_n} \subseteq \mathcal{R}_r^G \ \forall i, j$. A less trivial fact, which follows from more general results about Cohen-Macaulay rings ([HE], Proposition 13) is that $\mathcal{R}_r^G$ is actually a free module of finite rank over the subalgebra generated by these $\{e_i(x^{(j)})\}_{\substack{i=1,\ldots,n \\ j=1,\ldots,r}}$. Thus there exist $\eta_1, \ldots, \eta_t \in \mathcal{R}_r^G$ such that any $f \in \mathcal{R}_r^G$ can be written uniquely in the form

$$f = \sum_{l=1}^t \eta_l p_l(e_i(x^{(j)}))$$

where each p_l is some polynomial in rn variables with coefficients in $\mathbf{Q}$. The nicest description of $\mathcal{R}_r^G$ that we will seek is an explicit choice of such a basis $\eta_1, \ldots, \eta_t$.

Garsia and Stanton ([GS]) examined this problem for the case $r = 1$. Their approach was to introduce a different ring $\mathcal{Q}$ having an S_n-action such that nice descriptions for $\mathcal{Q}^G$ yield the same for $\mathcal{R}_r^G = \mathcal{R}_1^G$. We introduce an analogous ring $\mathcal{Q}_r$ for the general case.

Definition: Let $\mathcal{B}_n$ denote the *Boolean algebra* of rank n, i.e. the poset of all subsets of $\{1, 2, \ldots, n\}$ ordered under inclusion. Let $\mathcal{Q} = \mathbf{Q}[\mathcal{B}_n - \hat{0}]$ be the Stanley-Reisner ring of $\mathcal{B}_n - \hat{0}$, i.e.

$$\mathcal{Q} = \mathbf{Q}[y_J : \emptyset \neq J \subseteq \{1, 2, \ldots, n\}\,]/(y_J y_K : J \not\subseteq K, K \not\subseteq J).$$

Let

$$\mathcal{Q}_r \;=\; \mathcal{Q}\otimes\cdots\otimes\mathcal{Q} \;=\; \mathbf{Q}[y_J^{(j)} : \emptyset \neq J \subseteq \{1,2,\ldots,n\}\,]/(y_J^{(j)} y_K^{(j)} : J \not\subseteq K, K \not\subseteq J).$$

Define an $\mathbf{N}^{nr}$-grading on $\mathcal{Q}_r$ by setting $\deg(j_J^{(j)}) = \epsilon_{\#J,j}$, where $\epsilon_{i,j}$ is the $(i,j)^{th}$ standard basis vector in $\mathbf{N}^{nr}$. Define an S_n-action on $\mathcal{Q}_r$ by $\sigma(y_J^{(j)}) = y_{\sigma(J)}^{(j)}$, and note that this action preserves the grading.

As $\mathbf{Q}$-vector spaces with S_n-actions, $\mathcal{Q}_r$ and $\mathcal{R}_r$ are closely related.

Definition: The *transfer map* $T : \mathcal{Q}_r \rightarrow \mathcal{R}_r$ is defined by first setting $T(y_J^{(j)}) = \prod_{i\in J} x_I$, then extending multiplicatively on *non-zero* monomials $y_{J_1}^{(j_1)} \cdots y_{J_l}^{(j_l)}$ (this monomial is non-zero in $\mathcal{Q}_r$ if $j_m = j_n$ implies either $J_m \subseteq J_n$ or $J_n \subseteq J_m$), and then extending $\mathbf{Q}$-linearly to all of $\mathcal{Q}_r$.

Define the *rank-row polynomials*

$$\theta_i^{(j)} = \sum_{\substack{J \subseteq \{1,\ldots,n\} \\ \# J = i}} y_J^{(i)}$$

for $i = 1,\ldots,n$ and $j = 1,\ldots,r$, and note that $T(\theta_i^{(j)}) = e_i(x^{(j)})$.

Example: Let $n = 3, r = 2$. Then

$$T(y_3^{(1)} \cdot y_3^{(1)} \cdot y_{123}^{(1)} \cdot y_2^{(2)} \cdot y_{23}^{(2)} + y_1^{(1)} \cdot y_{12}^{(1)}) =$$

$$x_3^{(1)} \cdot x_3^{(1)} \cdot x_1^{(1)} x_2^{(1)} x_3^{(1)} \cdot x_2^{(2)} \cdot x_2^{(2)} x_3^{(2)} + x_1^{(1)} \cdot x_1^{(1)} x_2^{(2)} =$$

$$(x_3^{(1)})^3 x_1^{(1)} x_2^{(1)} (x_2^{(2)})^2 x_3^{(2)} + (x_1^{(1)})^2 x_2^{(2)}.$$

It is an easy exercise (or see [Ga], Section 6) to show that $T : \mathcal{Q}_1 \rightarrow \mathcal{R}_1$ is a $\mathbf{Q}$-linear isomorphism, and hence that $T : \mathcal{Q}_r \rightarrow \mathcal{R}_r$ is also. Furthermore, since T commutes with the S_n-actions on $\mathcal{Q}_r$ and $\mathcal{R}_r$, this implies $T : \mathcal{Q}_r^G \rightarrow \mathcal{R}_r^G$ is a $\mathbf{Q}$-linear isomorphism also. This yields the following:

Proposition 4.2.1 *Let*

$$F(Q_r, \lambda) = \sum_{\alpha \in \mathbf{N}^{nr}} dim_{\mathbf{Q}}(Q_r^G)_\alpha \lambda^\alpha$$

(where $\lambda^\alpha = \prod_{i=1}^{n} \prod_{j=1}^{r} (\lambda_i^{(j)})^{\alpha_{i,j}}$ if $\alpha = \sum \alpha_{i,j} \epsilon_{i,j} \in \mathbf{N}^{nr}$) be the Hilbert series for Q_r^G. Then

$$F(\mathcal{R}_r^G, t) = F(Q_r^G, \lambda)|_{\lambda_i^{(j)} \mapsto (t^{(j)})^i}.$$

Proof: Note that $y_J^{(j)}$ is counted as $\lambda_{\#J}^{(j)}$ in $F(Q_r, \lambda)$, while $T(y_J^{(j)})$ is counted as $(t^{(j)})^{\#J}$ in $F(\mathcal{R}_r^G, t)$. Since T preserves the grading in this fashion, and is a **Q**-linear isoomorphism, the result follows.∎

We would like then to compute $F(Q_r^G, \lambda)$. To do this, we follow the lead of [GS], by relating Q_r and $\Sigma((S_n)^r, rS)$ explicitly.

Note first that $Q_r = Q_r'[y_{12\cdots n}^{(j)} : j = 1, \ldots, n]$ where

$$Q_r' = \mathbf{Q}[y_J^{(j)} : \emptyset \neq J \subset \{1, 2, \ldots, n\}, j = 1, \ldots, n]$$

and that $y_{12\cdots n}^{(j)} \in Q_r^G$ for all j, so

$$Q_r^G = Q_r'^G[y_{12\cdots n} : j = 1, \ldots, n].$$

Proposition 4.2.2

$$Q_r' = \mathbf{Q}[\Sigma((S_n)^r, rS)] = \textit{the face ring of } \Sigma((S_n)^r, rS)$$

Proof: Recall that $\Sigma(S_n, S)$ is the barycentric subdivision of the boundary of $(n-1)$-simplex having vertices $\{1, 2, \ldots, n\}$. Thus the vertices of $\Sigma(S_n, S)$ may be identified with subsets $\emptyset \neq J \subset \{1, 2, \ldots, n\}$, and faces of $\Sigma(S_n, S)$ may be identified with chains of such subsets. But this means $\mathbf{Q}[\Sigma(S_n, S)] = Q_1'$, by definition of Q_1'. Clearly, $Q_r' = Q_1' \otimes \cdots \otimes Q_1'$, and we have already noted that $\Sigma((S_n)^r, rS) = \Sigma(S_n, S) * \cdots * \Sigma(S_n, S)$. Thus by the fact that $k[A * B] = k[A] \otimes k[B]$ for any two simplicial complexes A, B, the result follows.∎

It is important to keep track of the correspondence between the two labellings we are implicitly using for $\Sigma((S_n)^r, rS)$ and $\Sigma(S_n, S)$. On the one hand, a vertex of $\Sigma(S_n, S)$ thought of as a coset wW_{S-s} has the label $s \in \{(12), \ldots, (n-1\ n)\}$. On the other hand, a vertex thought of as a subset $\emptyset \neq K \subset \{1, 2, \ldots, n\}$ has the label $\#K \in \{1, 2, \ldots, n\}$. By chasing through the definitions, one can check that if $s = (i\ i+1)$, then wW_{S-s} is a vertex corresponding to a subset L with $\#L = i$. This labelling correspondence extends straightforwardly to $\Sigma((S_n)^r, rS)$. Finally, we can state:

Proposition 4.2.3

$$F(\mathcal{Q}_r^G, \lambda) =$$

$$\frac{1}{\prod_{i=1}^n \prod_{j=1}^r (1 - \lambda_i^{(j)})} \sum_{J_1, \ldots, J_r \subseteq S} \beta_{J_1, \ldots, J_r}(\Sigma((S_n)^r, rS)/\Delta^r(G)) \prod_{s=1}^r \prod_{l \in J_s} \lambda_l^{(s)}$$

Proof: ¿From the relation between $\mathcal{Q}_r^G$ and $\mathcal{Q}_r'^G$, we have

$$F(\mathcal{Q}_r^G, \lambda) = \frac{1}{\prod_{j=1}^r (1 - \lambda_n^{(j)})} F(\mathcal{Q}_r'^G, \lambda).$$

A **Q**-basis for $\mathcal{Q}_r'^G$ is in one-to-one corrspondence with G-orbits Gm of non-zero monomials $m \in \mathcal{Q}_r'$. Thus we have

$$F(\mathcal{Q}_r'^G, \lambda) = \sum_{Gm} \lambda^{deg(m)}.$$

Given a monomial $m = \prod_i (y_{J_i}^{(j_i)})^{m_i}$, we will say its *support* is the square-free monomial $\operatorname{supp}(m) = \prod_i y_{J_i}^{(j_i)}$. Notice that two monomials m, m' are in the same $G - orbit$ if and only if

$$G\operatorname{supp}(m) = G\operatorname{supp}(m') \text{ and } \deg(m) = \deg(m').$$

Thus we have

$$F(Q_r'^G, \lambda)$$

$$= \sum_{\substack{Gm \\ m\ square-free}} \sum_{\substack{Gm' \\ Gsupp(m')=Gm}} \lambda^{deg(m)}$$

$$= \sum_{\substack{Gm: \\ m\ square-free}} \frac{\lambda^{deg(m)}}{\prod_{i,j}(1-\lambda_i^{(j)})^{deg(m)_{i,j}}}$$

$$= \sum_{J_1,\ldots,J_r \subseteq \{1,\ldots,n-1\}} \# \left\{ Gm : \lambda^{deg(m)} = \prod_{s=1}^r \prod_{l\in J_i} \lambda_l^{(s)} \right\} \cdot \frac{\prod_{s=1}^r \prod_{l\in J_i} \lambda_l^{(s)}}{\prod_{s=1}^r \prod_{l\in J_i} (1-\lambda_l^{(s)})}$$

Now if we convert the label sets $J_i \subseteq \{1,\ldots,n-1\}$ into subsets $J_i \subseteq S$ (using the scheme discussed above), then the previous proposition implies

$$\#\{Gm : \lambda^{deg(m)} = \prod_{s=1}^r \prod_{l\in J_i} \lambda_l^{(s)}\} = \alpha_{J_1,\ldots,J_r}(\Sigma((S_n)^r, rS)/\Delta^r(G))$$

and hence

$$F(Q_r'^G, \lambda) =$$

$$\sum_{J_1,\ldots,J_r \subseteq \{1,\ldots,n-1\}} \alpha_{J_1,\ldots,J_r}(\Sigma((S_n)^r, rS)/\Delta^r(G)) \frac{\prod_{s=1}^r \prod_{l\in J_i} \lambda_l^{(s)}}{\prod_{s=1}^r \prod_{l\in J_i} (1-\lambda_l^{(s)})}.$$

Bringing this over a common denominator (and a little algebra) gives

$$F(Q_r'^G, \lambda) =$$

$$\frac{1}{\prod_{i=1}^{n-1} \prod_{j=1}^r (1-\lambda_i^{(j)})} \sum_{J_1,\ldots,J_r \subseteq S} \beta_{J_1,\ldots,J_r}(\Sigma((S_n)^r, rS)/\Delta^r(G)) \prod_{s=1}^r \prod_{l\in J_s} \lambda_l^{(s)}$$

and combining this with the first sentence of this proof gives the result. ∎

Corollary 4.2.4 *Let $W' \subseteq S_n$ be a reflection subgroup. Then*

$$F(Q_r^{W'}, \lambda) = \frac{1}{\prod_{i=1}^n \prod_{j=1}^r (1-\lambda_i^{(j)})} \sum_{\substack{(\sigma_1,\ldots,\sigma_r)\in S_n^r \\ I(\sigma_r\cdots\sigma_1)\cap W'=\emptyset}} \prod_{s=1}^r \prod_{l\in D(\sigma_i)} \lambda_l^{(s)}$$

and

$$F(R_r^{W'}, \lambda) = \frac{1}{\prod_{i=1}^n \prod_{j=1}^r (1-(t^{(j)})^i)} \sum_{\substack{(\sigma_1,\ldots,\sigma_r)\in S_n^r \\ I(\sigma_r\cdots\sigma_1)\cap W'=\emptyset}} \prod_{s=1}^r (t^{(s)})^{maj(\sigma_i)}$$

where $maj(\sigma) = \sum_{(i\ i+1)\in D(\sigma)} i$ is called the major (or greater) index *of the permutation σ.* ∎

Having found one of our "nice descriptions" of $\mathcal{R}_r^G$, we now look at the other. The following two theorems may be proven as straighforward extensions of the analogous results for the $r = 1$ case given in [GS].

Theorem 4.2.5 (cf. [GS], Theorem 9.2) *If $\gamma_1, \ldots, \gamma_t \in \mathcal{Q}_r^G$ are homogenous and form a basis for $\mathcal{Q}_r^G$ as a free module over the subalgebra $\mathbf{Q}[\theta_i^{(j)}]_{\substack{i=1,\ldots,n \\ j=1,\ldots,r}}$, then*

$$\eta_1 = T(\gamma_1), \ldots, \eta_t = T(\gamma_t)$$

form a basis for $\mathcal{R}_r^G$ as a free module over the subalgebra

$$\mathbf{Q}[e_i(x^{(j)})]_{\substack{i=1,\ldots,n \\ j=1,\ldots,r}} \text{ .} ∎$$

Theorem 4.2.6 (cf. [GS], Theorem 6.2) *Let*

$$\Sigma((S_n)^r, rS)/\Delta^r(G) = \coprod_{i=1}^{t} [F_i, M_i]$$

be a shelling, and for $F \in \Sigma((S_n)^r, rS)/\Delta^r(G)$ let

$$\mathcal{S}^G(F) = \frac{1}{\#G} \sum_{m' \in Gm} m'$$

where Gm is the orbit of monomials in $\mathcal{Q}_r$ corresponding to the orbit of faces F. Then $\gamma_1 = \mathcal{S}^G(F_1), \ldots, \gamma_t = \mathcal{S}^G(F_t)$ form a basis as in the hypothesis of the previous theorem.

We are now but a definition away from our goal.

Definition: Given $J = \{(i_1\ i_1 + 1), \ldots, (i_l\ i_l + 1)\} \subseteq S$ and $\sigma \in S_n$, let

$$\gamma_J^{(j)}(\sigma) = y_{\sigma_1 \cdots \sigma_{i_1}}^{j}\, y_{\sigma_1 \cdots \sigma_{i_2}}^{j} \cdots y_{\sigma_1 \cdots \sigma_{i_l}}^{j} \in \mathcal{Q}_r$$

and let

$$\gamma_{J_1,\ldots,J_r}(\sigma_1,\ldots,\sigma_r) = \gamma_{J_1}^{(j)}(\sigma_1)\cdots\gamma_{J_r}^{(j)}(\sigma_r)$$

and

$$\eta_{J_1,\ldots,J_r}(\sigma_1,\ldots,\sigma_r) = T(\gamma_{J_1,\ldots,J_r}(\sigma_1,\ldots,\sigma_r)).$$

Let $\mathcal{S}^G$ denote the symmetrization operator defined in the previous theorem.

Theorem 4.2.7 *For $r = 1,2$ the set*

$$\{\mathcal{S}^{W'}\gamma_{D(w_1),\ldots,D(w_r)}(\sigma_r\sigma_{r-1}\cdots\sigma_1,\sigma_r\sigma_{r-1}\cdots\sigma_2,\ldots,\sigma_r):$$
$$I(w_r w_{r-1}\cdots w_1)\cap W' = \emptyset\}$$

form a basis as in Theorem 4.2.5 for $\mathcal{Q}_r^{W'}$, and hence their images under T (the corresponding η's) form a basis for $\mathcal{R}_r^{W'}$.

Proof: Since Theorem 4.1.6 gives a shelling of $\Sigma(W^r, rS)/\Delta^r(W')$ for $r = 1, 2$, we can apply Theorems 4.2.5 and 4.2.6, yielding the result.∎

Example: Let $n = 3, r = 2, W' = W = S_n$. Writing the permutation $\sigma = \begin{pmatrix} 1 & 2 & 3 \\ \sigma_1 & \sigma_2 & \sigma_3 \end{pmatrix}$ as $\sigma_1\sigma_2\sigma_3$, and putting a dot between σ_i, σ_{i+1} iff $(i \; i+1) \in D(\sigma)$, we have the following table:

(w_1,w_2): $I(w_2w_1)\cap W'=\emptyset$	(w_2w_1, w_2)	$\mathcal{S}^{W'}\gamma_{D(w_1).D(w_2)}(w_2w1, w_2)$	$\mathcal{S}^{W'}\eta_{D(w_1),D(w_2)}(w_2w1, w_2)$
$(123, 123)$	$(123, 123)$	$\mathcal{S}^{W'}1$	$\mathcal{S}^{W'}1$
$(13\cdot2, 13\cdot2)$	$(123, 132)$	$\mathcal{S}^{W'}y_{12}^{(1)}y_{13}^{(2)}$	$\mathcal{S}^{W'}x_1^{(1)}x_2^{(1)}x_1^{(2)}x_3^{(2)}$
$(2\cdot13, 2\cdot13)$	$(123, 213)$	$\mathcal{S}^{W'}y_1^{(1)}y_2^{(2)}$	$\mathcal{S}^{W'}x_1^{(1)}x_2^{(2)}$
$(23\cdot1, 3\cdot12)$	$(123, 312)$	$\mathcal{S}^{W'}y_{12}^{(1)}y_3^{(2)}$	$\mathcal{S}^{W'}x_1^{(1)}x_2^{(1)}x_3^{(2)}$
$(3\cdot12, 23\cdot1)$	$(123, 132)$	$\mathcal{S}^{W'}y_1^{(1)}y_{23}^{(2)}$	$\mathcal{S}^{W'}x_1^{(1)}x_2^{(2)}x_3^{(2)}$
$(3\cdot2\cdot1, 3\cdot2\cdot1)$	$(123, 321)$	$\mathcal{S}^{W'}y_1^{(1)}y_{12}^{(1)}y_3^{(2)}y_{23}^{(2)}$	$\mathcal{S}^{W'}x_1^{(1)}x_1^{(1)}x_2^{(1)}x_3^{(2)}x_2^{(2)}x_3^{(2)}$

¿From the previous theorem, we conclude that the symmetrized monomials in the third column form a basis for $\mathcal{Q}_2^{S_3}$ as a free module over $\mathbf{Q}[\theta_i^{(j)}]_{\substack{j=1,2 \\ i=1,2,3}}$, and those in the fourth column form a basis for $\mathcal{R}_2^{S_3}$ as a free module over $\mathbf{Q}[e_i(x^{(j)})]_{\substack{j=1,2 \\ i=1,2,3}}$. Notice also that the data about descents shown in the first column verifies an instance of Corollary 4.1.5.

Conjecture 4.2.8 *Theorem 4.2.7 holds without the restriction to $r = 1, 2$.*

This conjecture cannot be proven in general by appeal to Theorem 4.2.6, since we have seen in an earlier remark that $\Sigma((S_n)^r, rS)/\Delta^r(W')$ may be non-shellable. However, Garsia and Stanton prove for $r = 1$ (and it easily generalizes to all r), that the conclusion to Theorem 4.2.7 *is equivalent to* the weaker hypothesis that $\Sigma((S_n)^r, rS)/\Delta^r(W') = \coprod_{i=1}^t [F_i, M_i]$ is a partitioning for which the *incidence matrix*

$$(m_{i,j})_{i,j=1}^t \text{ where } m_{i,j} = \begin{cases} 1 \text{ if } F_i \leq M_i \\ 0 \text{ else} \end{cases}$$

is invertible. It is clear that $\coprod_{i=1}^t [F_i, M_i]$ is a shelling exactly when $(m_{i,j})$ is upper triangular (and hence invertible, since $m_{i,i} = 1$).

Admittedly, the evidence in support of the invertibility of $(m_{i,j})$ for $r \geq 3$ (and hence for the above conjecture) is small, since there are only two special cases for which we can prove it:

1. For $W' = 1$, since in this case $\Sigma((S_n)^r, rS)/\Delta^r(W') = \Sigma((S_n)^r, rS)$, and then the above partitioning is the same as the shelling of Theorem 3.4.5.

2. For $n = 2$ and $W' = W = S_2$, by an ad hoc induction on r (which we mercifully omit).

Remark: For $r = 1, 2$ and $W' = W = S_n$, Theorem 4.2.7 gives an explicit description of a ring considered by Solomon in his invariant-theoretic proof of Gordon's Theorem. To be precise, in the notation of Theorem 4.12 of [So3], if we take $G = 1, W = \mathbf{C}, p = 1$, and $Y_1 = X_1$, then Theorem 4.12 (Gordon's Theorem) asserts that a certain formal power series P_n has non-negative integral coefficients. The proof proceeds by showing that P_n is the Hilbert series of a ring which Solomon calls $I(T^m C^n \otimes W^n)$. With the above choices for G, W, p, Y_1, this ring is the same as (in our notation) the ring

$$\mathcal{R}_n^{S_n} / (e_i(x^{(j)}))_{\substack{i=1,\dots,n \\ j=1,\dots,m}} .$$

Thus Theorem 4.2.7 gives a **Q**-basis for this ring if $n = 1, 2$, and the succeeding conjecture asserts the same for all n.

Remark: All the results of this section have analogues for *Weyl groups W* other than S_n . For information on this, see [GS], Sections 8,9.

4.3 Alternating subgroups and their diagonal embeddings

In this section we examine quotients by another class of subgroups, related to reflection subgroups.

Definition: Let W' be a reflection subgroup of W. The *alternating subgroup*

E' of W' is defined by $E' = \{w \in W : \text{sgn}(w) = 1\}$. For example, if $W' = W = S_n$, then E' is the subgroup of all even permutations in S_n .

For the remainder of this section, let (W, S) be a finite Coxter system, W' a reflection subgroup of W, and E' the alternating subgroup of W'. We now propose to study quotients $\Sigma(W^r, rS)/\Delta^r(E')$, just as we did for $\Sigma(W^r, rS)/\Delta^r(W')$ in Section 4.1. We need one more piece of Coxeter group theory before we can proceed.

Definition: The *longest element* w_0 of W is the unique element of W satisfying $I(w_0) = \Phi^+$ (i.e. $w_0 \Phi^+ = -\Phi^+$). We will also need the fact that $w_0^2 = 1$ (see [Bo], Chapitre VI Section 1, Corollaire 3 for facts about w_0). Since W' is a Coxeter group in its own right, it also has a unique longest element (which we will call w_0') satisfying $I(w_0') \cap W' = \Phi_{W'}^+$. Of course, we also have $w_0'^2 = 1$.

Example: For $W = S_n$, $w_0 = \begin{pmatrix} 1 & 2 & \cdots & n \\ n & n-1 & \cdots & 1 \end{pmatrix}$. For $W' \subseteq W = S_n$, if W' corresponds to the partition π of $\{1, 2, \ldots, n\}$, then w_0' is the unique permutation in W' that has the numbers in each block of π in decreasing order. E.g., if $n = 6$ and $W' = S_{\{1,4,5\}} \times S_{\{2,6\}} \times S_{\{3\}}$ then $w_0' = \begin{pmatrix} 123456 \\ 563412 \end{pmatrix}$.

For the remainder of this section, we fix a particular reflection t in W' (i.e $t \in W' \cap T$).

Proposition 4.3.1

$$\mathcal{A}_r(P(W')) \ \amalg \ tw_0'\mathcal{A}_r(-P(W'))$$

is a fundamental domain for the action of E' on V^r, i.e. every orbit $E'f$ of a vector $f \in V^r$ has a unique representative in the above set.

Proof: First we show that the above union is indeed disjoint. Suppose not, i.e. let $e \in \mathcal{A}_r(P(W')) \cap tw_0'\mathcal{A}_r(-P(W'))$. Let $t = r_\beta$ with $\beta \in \Phi_{W'}^+$, and let $\alpha = w_0'(\beta) \in -\Phi_{W'}^+$ (since $w_0'\Phi_{W'}^+ = -\Phi_{W'}^+$). Then we have

$$\langle \beta, e \rangle = -\langle t(\beta), e \rangle = -\langle tw_0'w_0'(\beta), e \rangle = -\langle tw_0'(\alpha), e \rangle = -\langle \alpha, w_0't(e) \rangle <_{\mathcal{L}} \underline{0}$$

since $e \in tw_0'\mathcal{A}_r(-P(W'))$ and $\alpha \in -P(W') \cap -\Phi^+$. But $\langle \beta, e \rangle <_{\mathcal{L}} \underline{0}$ contradicts $e \in \mathcal{A}_r(P(W'))$, since $\beta \in \Phi_{W'}^+$.

Existence: Given $E'f$, let e' be the unique representative of $W'f$ which lies in $\mathcal{A}_r(P(W'))$ (and whose existence is guaranteed by Proposition 4.1.1). If $e' \in E'f$, then let $e = e'$ and we are done. Otherwise $e' \in W'f - E'f$, so let $e = t(e') \in E'f$. We claim that in this case, $e \in tw_0'\mathcal{A}_r(-P(W'))$, i.e. $w_0't(e) \in \mathcal{A}_r(-P(W'))$. To see this, let $\alpha \in -\Phi_{W'}^+$, and we have

$$\langle \alpha, w_0't(e) \rangle = \langle \alpha, w_0'(e') \rangle = \langle w_0'(\alpha), e' \rangle \geq_{\mathcal{L}} \underline{0}$$

since $w_0'(\alpha) \in \Phi_{W'}^+$ and $e' \in \mathcal{A}_r(-P(W'))$. We actually need the previous inequality to be strict. But if it were not strict, that is if $\langle w_0'(\alpha), e' \rangle = \underline{0}$, then $e' = r_{w_0'(\alpha)}(e') \in E'f$, a contradiction.

Uniqueness: Let $e_1, e_2 \in E'f$ both lie in $\mathcal{A}_r(P(W')) \amalg tw_0'\mathcal{A}_r(-P(W'))$. We must show $e_1 = e_2$.

Case 1: $e_i \in \mathcal{A}_r(P(W'))$ for $i = 1, 2$. Then $e_1 = e_2$ by the uniqueness statement in Proposition 4.1.1.

Case 2: $e_i \in tw_0'\mathcal{A}_r(-P(W'))$. It is easy to check that

$$e \in tw_0'\mathcal{A}_r(-P(W')) \Rightarrow t(e) \in \mathcal{A}_r(P(W')).$$

Hence in this case we have $t(e_1), t(e_2) \in \mathcal{A}_r(P(W')) \cap W'f$, so $t(e_1) = t(e_2)$ and $e_1 = e_2$.

Case 3: $e_1 \in \mathcal{A}_r(P(W'))$, $e_2 \in tw'_0 \mathcal{A}_r(-P(W'))$. We want show that this leads to a contradiction. Since $e_1, t(e_2) \in \mathcal{A}_r(P(W')) \cap W'f$, we have $e_1 = t(e_2)$. On the other hand, since $e_1, e_2 \in E'f$, we have $e_1 = \epsilon(e_2)$ for some $\epsilon \in E'$. We will get our contradiction by showing that $t = \epsilon$ (impossible since $t \notin E'$). To see this, let $V_{W'}$ be the **R**-span of $\Phi^+_{W'}$, and let $\pi : V \to V_{W'}$ be orthogonal projection with respect to $\langle \cdot, \cdot \rangle$. Given $\alpha \in \Phi^+_{W'}$, we have

$$\langle \alpha, e_1 \rangle = \langle \alpha, t(e_2) \rangle = \langle \alpha, w'_0 w'_0 t(e_2) \rangle = \langle w'_0(\alpha), w'_0 t(e_2) \rangle >_{\mathcal{L}} \underline{0}$$

since $w'_0(\alpha) \in -\Phi^+_{W'}$ and $w'_0 t(e_2) \in \mathcal{A}_r(-P(W'))$. Thus

$$\langle \alpha, \pi(e_1) \rangle = \langle \alpha, \pi(t(e_2)) \rangle >_{\mathcal{L}} \underline{0} \; \forall \alpha \in \Phi^+_{W'},$$

and since W' preserves $V_{W'}$ and commutes with π, this means

$$\langle \alpha, \epsilon\pi(e_2) \rangle = \langle \alpha, t\pi(e_2) \rangle >_{\mathcal{L}} \underline{0} \; \forall \alpha \in \Phi^+_{W'}.$$

But $\Phi^+_{W'}$ is a positive root system for W' on $V_{W'}$ (Appendix, Proposition A.0.8), and hence by Lemma 4.1.2, we have $\epsilon = t$.∎

Example: The previous proposition is easy to understand when $W = S_n$, and particularly simple when $W' = W = S_n$ and $r = 1$ (although the more general case of W' and r is very similar). In this case, the proposition says that given $(f_1, \ldots, f_n) \in \mathbf{R}^n$, we can either use an even permutation ϵ to get $f_{\epsilon(1)} \geq \ldots \geq f_{\epsilon(1)}$ (i.e. $\epsilon(f) \in \mathcal{A}(P(W'))$), or else this is impossible. If it is impossible, then all of the f_i's must be distinct, and by an odd permutation σ we can get $f_{\sigma(1)} > \ldots > f_{\sigma(n)}$. Hence if we fix $t = (12)$, then using the even permutation $t\sigma$ we can get

$$f_{t\sigma(2)} > f_{t\sigma(1)} > f_{t\sigma(3)} > f_{t\sigma(4)} \cdots > f_{t\sigma(n)}$$

i.e. $\sigma t(f) \in t w_0 \mathcal{A}(-P(W'))$.

Theorem 4.3.2

$$\Sigma(W^r, rS)/\Delta^r(E') \; =$$

$$\coprod_{\substack{(w_1,\ldots,w_r) \\ I(w_r\cdots w_1)\cap W'=\emptyset}} \Delta^r(E') \prod_{i=1}^r [w_r w_{r-1}\cdots w_i W_{S-D(w_i)}, w_r w_{r-1}\cdots w_i W_\emptyset]$$

$$\coprod_{\substack{(w_1,\ldots,w_r) \\ I(w_r\cdots w_1)\cap W'=\Phi_{W'}^+}} \Delta^r(E') \prod_{i=1}^r [t w_0' w_r w_{r-1}\cdots w_i W_{S-D(w_i)}, t w_0' w_r w_{r-1}\cdots w_i W_\emptyset]$$

is a partitioning.

Proof: The previous proposition asserts that

$$V^r/E' = \{E'f : f \in \mathcal{A}_r(P(W')) \text{ II } t w_0' \mathcal{A}_r(-P(W')) \}.$$

Since

$$\mathcal{L}(-P(W')) \; = \{w \in W : -\Phi_{W'}^+ \subseteq w\Phi^+\} = \{w \in W : I(w) \cap W' = \Phi_{W'}^+\},$$

we conclude from Proposition 3.4.1 that

$$V^r/E' = \{E'f : f \in \coprod_{w:I(w)\cap W'=\emptyset} \mathcal{A}_r(w\Phi^+) \text{ II } \coprod_{w:I(w)\cap W'=\Phi_{W'}^+} t w_0' \mathcal{A}_r(w\Phi^+) \}.$$

Proceeding as in the proof of Theorem 4.1.3, we apply Theorem 3.4.2, then apply the map $E'f \mapsto F(E'f)$, and use Lemma 3.2.1 to reach our conclusion.∎

Corollary 4.3.3

$$\beta_{J_1,\ldots,J_r}(\Sigma(W^r, rS)/\Delta^r(E')\,) = \#\{(w_1,\ldots,w_r): I(w_r\cdots w_1)\cap W' = \emptyset \text{ or } \Phi_{W'}^+\}.∎$$

Corollary 4.3.4

$$\#\{(w_1, \ldots, w_r) : D(w_i) = J_i, I(w_r \cdots w_1) \cap W' = \emptyset \text{ or } \Phi^+_{W'},\} =$$

$$\#\{(w_1, \ldots, w_r) : D(w_i) = S - J_i, I(w_r \cdots w_1) \cap W' = \emptyset \text{ or } \Phi^+_{W'}\}$$

for all $J_1, \ldots, J_r \subseteq S$.

Proof: By definition, $\mathrm{sgn}(\epsilon) = 1$ for all $\epsilon \in E'$. Apply Proposition 2.4.4 and the previous corollary.■

Remark: The exact same remarks as after Corollary 4.1.5 apply to the previous corollary.

When $r = 1$, we can put a shelling order on the above partitioning. But prior to doing this, let us to write the partitioning more succinctly. Note that

$$
\begin{aligned}
u \in \mathcal{L}(-P(W')) \quad &\Leftrightarrow \quad -\Phi^+_{W'} \subseteq u\Phi^+ \\
&\Leftrightarrow \quad w_0' \Phi^+_{W'} \subseteq u\Phi^+ \\
&\Leftrightarrow \quad \Phi^+_{W'} \subseteq w_0' u\Phi^+ \\
&\Leftrightarrow \quad I(w_0' u) \cap W' = \emptyset \\
&\Leftrightarrow \quad u \in w_0' \mathcal{L}(P(W'))
\end{aligned}
$$

Thus for $r = 1$ we can rewrite our partitioning as

$$
\begin{aligned}
\Sigma(W,S)/E' &= \coprod_{w \in \mathcal{L}(P(W'))} [E'wW_{S-D(w)}, E'wW_\emptyset] \amalg [E'twW_{S-D(w_0'w)}, E'twW_\emptyset] \\
&= \coprod_{w \in \mathcal{L}(P(W')) \amalg t\mathcal{L}(P(W'))} [E'uW_{S-D(\psi(w))}, E'uW_\emptyset]
\end{aligned}
$$

where $\psi : \mathcal{L}(P(W')) \amalg t\mathcal{L}(P(W')) \to W$ is the set map defined by

$$\psi(u) = \begin{cases} u \text{ if } u \in \mathcal{L}(P(W')) \\ w_0' tu \text{ if } u \in t\mathcal{L}(P(W')) \end{cases}$$

Theorem 4.3.5 *For $r = 1$, the above partitioning is a shelling, if we order*

$$\{u \in \mathcal{L}(P(W')) \amalg t\mathcal{L}(P(W'))\}$$

by any linear extension of Bruhat order $<_B$ on $\{\psi(u)\}_{u \in \mathcal{L}(P(W')) \amalg t\mathcal{L}(P(W'))}$.

Proof: We need to show that if

$$u_1, u_2 \in \mathcal{L}(P(W')) \amalg t\mathcal{L}(P(W'))$$

and

$$E' u_2 \subseteq E' u_1 W_{S-D(\psi(u_1))},$$

then $\psi(u_1) \leq_B \psi(u_2)$. There are four cases:

Case 1: $u_1, u_2 \in \mathcal{L}(P(W'))$. Then we have $E' u_2 \subseteq E' u_1 W_{S-D(u_1)}$ which implies $u_2 \in W' u_1 W_{S-D(u_1)}$. Hence $u_1 \leq_B u_2$, since u_1 is the least element of $W' u_1 W_{S-D(u_1)}$ under $\leq_B$ (Appendix, Proposition A.0.11). But $u_i = \psi(u_i)$ for $i = 1, 2$ and thus $\psi(u_1) \leq_B \psi(u_2)$.

Case 2: $u_1 \in \mathcal{L}(P(W'))$, $u_2 \in t\mathcal{L}(P(W'))$. Then we still have $u_2 \in W' u_1 W_{S-D(u_1)}$, and hence $w_0' u_2 \in W' u_1 W_{S-D(u_1)}$, so $u_1 \leq_B w_0' u_2$. But $u_1 = \psi(u_1), w_0' u_2 = \psi(u_2)$.

Case 3: $u_1, u_2 \in t\mathcal{L}(P(W'))$. Let $u_i = tv_i$ for $i = 1, 2$. Then $E' u_2 \subseteq E' u_1 W_{S-D(\psi(u_1))}$ implies $E' tv_2 \subseteq E' tv_1 W_{S-D(w_0' v_1)}$. Let $tv_2 \in etv_1 W_{S-D(v_1)}$ for some $e \in E'$. Then we have

$$te^{-1} tv_2 \in v_1 W_{S-D(w_0' v_1)} \Rightarrow w_0' te^{-1} tv_2 \in w_0' v_1 W_{S-D(w_0' v_1)} \Rightarrow w_0' v_1 \leq_B w_0' te^{-1} tv_2$$

since $w_0'v_1$ is the least element of $w_0'vW_{S-D(w_0'v)}$ under weak order $\leq_R$ and hence under $\leq_B$. We also have $w_0'te^{-1}t \leq_B w_0'$, since w_0' is the greatest element of W' under $\leq_B$. Hence $w_0'te^{-1}tv_2 \leq_B w_0'v_2$, since $I(v_2) \cap W' = \emptyset$ implies that multiplication of elements of W' on the right by v_2 preserves $\leq_B$ (Appendix, Proposition A.0.9). Thus

$$\psi(u_1) = w_0'v_1 \leq_B w_0'te^{-1}tv_2 \leq_B w_0'v_2 = \psi(u_2).$$

Case 4: $u_1 \in t\mathcal{L}(P(W'))$, $u_2 \in \mathcal{L}(P(W'))$. We will show that this leads to a contradiction. Let $u_1 = tv_1$. Then $E'u_2 \subseteq E'u_1W_{S-D(\psi(u_1))}$ implies $E'u_2 \subseteq E'tv_1W_{S-D(w_0'v_1)}$. Let $u_2 = etv_1w$ where $e \in E'$, $w \in W_{S-D(w_0'v_1)}$. We then have

$$
\begin{aligned}
u_2 &= etv_1w \\
te^{-1}u_2 &= v_1w \\
w_0'te^{-1}u_2 &= w_0'v_1w \\
I(w_0'te^{-1}u_2) &= I(w_0'v_1w)
\end{aligned}
$$

If we let $+$ denote the operation of symmetric difference of sets, then applying Lemma A.0.10 of the Appendix to the last equation, we get

$$I(w_0')+w_0'I(te^{-1})w_0'+w_0'te^{-1}I(u_2)etw_0' = (I(w_0')+w_0'I(v_1)w_0')\amalg w_0'v_1I(w)v_1^{-1}w_0'$$

If we intersect both sides of the above equation with W' and note that

$$I(v_1) \cap W' = I(u_2) \cap W' = \emptyset$$

implies

$$w_0'I(v_1)w_0' \cap W' = w_0'te^{-1}I(u_2)etw_0' \cap W' = \emptyset,$$

we get

$$T_{W'} - (w_0' I(te^{-1}) w_0' \cap W') = T_{W'} \text{ II } (w_0' v_1 I(e) v_1^{-1} w_0' \cap W')$$

This implies $w_0' I(te^{-1}) w_0 \cap W' = \emptyset$, and hence that $I(te^{-1}) \cap W' = \emptyset$, so $t = e$. But $t \notin E', e \in E'$, so this is a contradiction.∎

Corollary 4.3.6 $\Sigma(W,S)/E'$ is CM/k for all fields k and homeomorphic to an $(\#S - 1) - sphere$.

Proof: See the proof of Theorem 4.1.8.∎

Remark: Similarly to Theorem 4.1.6, one can give examples to show that Theorem 4.3.5 is tight in the sense that $\Sigma(W^r, rS)/\Delta^r(E')$ can be non-shellable for all $r \geq 2$. For example, let $(W, S) = (S_3, \{(12), (23)\}), W' = W = S_3, E' = \langle \binom{123}{231} \rangle \cong \mathbf{Z}_3$. It is easy to see that E' gives a *free* $\mathbf{Z}_3$-action on $\Sigma(W,S)$, and hence $\Delta^r(E')$ gives a free $\mathbf{Z}_3$-action on $\Sigma(W^r, rS)$ for all r. Hence for $r > 1$, since $\Sigma(W^r, rS)$ is a $(2r - 1)$-sphere and simply-connected, the quotient map $\Sigma(W^r, rS) \rightarrow \Sigma(W^r, rS)/\Delta^r(E')$ is the universal cover for $\Sigma(W^r, rS)/\Delta^r(E')$. We conclude that the fundamental group of $\Sigma(W^r, rS)/\Delta^r(E')$ is $E' \cong \mathbf{Z}_3$, and thus $\tilde{H}_1(\Sigma(W^r, rS)/\Delta^r(E')) = \mathbf{Z}_3$. Thus for $r \geq 2$, $\Sigma(W^r, rS)/\Delta^r(E')$ can not be shellable, since it is not CM/k for fields k of characteristic 3.

It is now a simple matter to apply the results (and notation) of Section 4.2 to prove the following results about invariants.

Theorem 4.3.7 Let W' be a reflection subgroup of S_n, and $E' \subseteq W'$ its *alternating subgroup. Then*

$$F(\mathcal{Q}_r^{E'}, \lambda) = \frac{1}{\prod_{i=1}^n \prod_{j=1}^r (1 - \lambda_i^{(j)})} \sum_{\substack{(\sigma_1,\ldots,\sigma_r) \in S_n^r \\ I(\sigma_r \cdots \sigma_1) \cap W' = \emptyset \ or \ \Phi_{W'}^+}} \prod_{s=1}^r \prod_{l \in D(\sigma_i)} \lambda_l^{(s)}$$

and

$$F(\mathcal{R}_r^{E'}, \lambda) = \frac{1}{\prod_{i=1}^n \prod_{j=1}^r (1 - (t^{(j)})^i)} \sum_{\substack{(\sigma_1,\ldots,\sigma_r) \in S_n^r \\ I(\sigma_r \cdots \sigma_1) \cap W' = \emptyset \ or \ \Phi_{W'}^+}} \prod_{s=1}^r (t^{(s)})^{maj(\sigma_i)}. \blacksquare$$

Theorem 4.3.8 *Let* W', E' *be as in the previous theorem. Then the set*

$$\{\mathcal{S}^{E'} \gamma_{D(\sigma)}(\sigma) : I(\sigma) \cap W' = \emptyset \ or \ \Phi_{W'}^+\}$$

form a basis as in Theorem 4.2.5 for $\mathcal{R}_r^{E'}$, *and hence their images under* T *(the corresponding* η's*) form a basis for* $\mathcal{R}_r^{E'}$. $\blacksquare$

Example: Let $(W, S) = (S_4, \{(12), (23), (34)\}), W' = \langle (13), (34) \rangle, E' = \langle (134) \rangle$. We know that $\Sigma(W, S)$ is the barycentric subdivision of the boundary of a tetrahedron with vertices labelled $1, 2, 3, 4$. In order to understand this particular E'-action, picture $\Sigma(W, S)$ as the suspension of a circle C, in which the circle C is the boundary of the triangle with corners $1, 3, 4$, and the suspension points are the vertices labelled 2 and 134 in the barycentric subdivision (see Figure 1). Since E' fixes both suspension points, and acts simplicially, it is not hard to see that in this case $\Sigma(W, S)/E' = \mathrm{Susp}(C)/E' = \mathrm{Susp}(C/E')$. It is easy to see that C/E' is a 1-sphere, and hence that $\Sigma(W, S)/E'$ is the suspension of this 1-sphere, i.e. a 2-sphere in accordance with Corollary 4.3.6.

If we choose $t = (13)$, then using the same notations as in the example of Section 4.2, we have the following table:

$w : I(w) \cap W' = \emptyset$	w	$\mathcal{S}^{E'} \gamma_{D(w)}(w)$	$\mathcal{S}^{E'} \gamma_{D(w)}(w)$
1234	1234	$\mathcal{S}^{E'} 1$	$\mathcal{S}^{E'} 1$
$2 \cdot 134$	2134	$\mathcal{S}^{E'} y_2$	$\mathcal{S}^{E'} x_2$
$13 \cdot 24$	1324	$\mathcal{S}^{E'} y_{13}$	$\mathcal{S}^{E'} x_1 x_3$
$134 \cdot 2$	1342	$\mathcal{S}^{E'} y_{134}$	$\mathcal{S}^{E'} x_1 x_3 x_4$

$w_0' w : I(w) \cap W' = \emptyset$	tw	$\mathcal{S}^{E'} \gamma_{D(w_0' w)}(tw)$	$\mathcal{S}^{E'} \gamma_{D(w_0' w)}(tw)$
$24 \cdot 3 \cdot 1$	2314	$\mathcal{S}^{E'} y_{23} y_{123}$	$\mathcal{S}^{E'} x_1 x_2^2 x_3$
$4 \cdot 23 \cdot 1$	3214	$\mathcal{S}^{E'} y_3 y_{123}$	$\mathcal{S}^{E'} x_1 x_2 x_3^2$
$4 \cdot 3 \cdot 12$	3142	$\mathcal{S}^{E'} y_3 y_{13}$	$\mathcal{S}^{E'} x_1 x_3^2$
$4 \cdot 3 \cdot 2 \cdot 1$	3124	$\mathcal{S}^{E'} y_3 y_{13} y_{123}$	$\mathcal{S}^{E'} x_1^2 x_2 x_3^2$

¿From the previous results, we have that the symmetrized monomials in the third column form a basis for $\mathcal{Q}^{E'}$ as a free module over $\mathbf{Q}[\theta_i]_{i=1,2,3,4}$, and those in the fourth column form a basis for $\mathcal{R}^{E'} = \mathbf{Q}[x_1, x_2, x_3, x_4]^{E'}$ as a free module over $\mathbf{Q}[e_i(x)]_{i=1,2,3,4}$. Notice also that the data about descents shown in the first column verifies an instance of Corollary 4.3.4

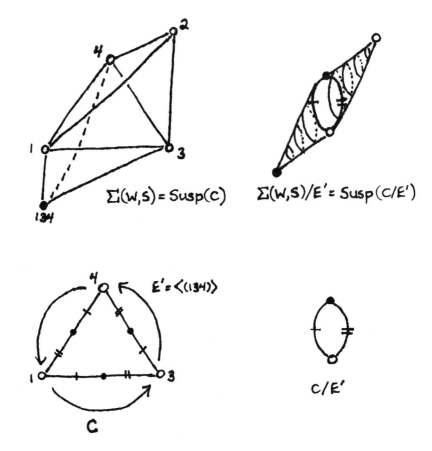

$$\Sigma(W,S) = \text{Susp}(C) \qquad \Sigma(W,S)/E' = \text{Susp}(C/E')$$

$$E' = \langle (134) \rangle$$

$$C$$

$$C/E'$$

Figure 4.1: An example of $\Sigma(W,S)/E'$

Chapter 5

Quotients by a Coxeter element

5.1 Introduction and definitions

In this chapter, we study quotients $\Sigma(W, S)/G$ for another class of subgroups G, namely cyclic subgroups generated by a Coxeter element.

Definition: Let (W, S) be a finite Coxeter system. We say $c \in W$ is a *Coxeter element* if $c = s_1 s_2 \cdots s_m$ for some ordering of $S = \{s_1, s_2, \ldots, s_m\}$. It is a fact (see [Bo], Chapitre V, Section 6) that for W finite, any two Coxeter elements are conjugate in W, i.e $s_1 \cdots s_m$ is conjugate in W to $s_{\sigma_1} \cdots s_{\sigma_m}$ for any permutation $\sigma \in S_m$. Thus for our purposes, we can fix one ordering of S and hence fix c for the remainder of the chapter. The *Coxeter number* h is defined to be the order of any Coxeter element, i.e. $h = \#\langle c \rangle$. The

93

exponents of W are defined to be the unique integers

$$1 \leq e_1 \leq e_2 \leq \ldots \leq e_m < h$$

such that $\{e^{\frac{2\pi i e_j}{h}}\}_{j=1,\ldots,m}$ are the eigenvalues of any Coxeter element c when c acts in the canonical representation of W as a reflection group.

Example: Let $(W,S) = (S_n \ , \{(12),(23),\ldots,(n-1\ n)\})$. Then

$$c = (12)(23)\cdots(n-1n) = (12\cdots n),$$

an n-cycle. Hence the Coxeter number h is n. To find the exponents, recall that in its canonical representation, S_n acts as permutations of the coordinates in $V = \{(f_1,\ldots,f_n) \in \mathbf{R}^n : \sum f_i = 0\}$. The characteristic polynomial for c acting on V is then $\frac{\lambda^n - 1}{\lambda - 1}$, so c's eigenvalues are the non-unit n^{th} roots of unity, and hence

$$\{e_j\}_{j=1,\ldots,m} = \{1,2,\ldots,n-1\}.$$

Definition: The Coxeter system (W,S) is *irreducible* if one cannot partition $S = S_1 \amalg S_2$ in such a way that every element of S_1 commutes with every element of S_2

Clearly every Coxeter system can be decomposed uniquely as

$$(W,S) = (W_1 \times \cdots \times W_r, S_1 \amalg \cdots \amalg S_r)$$

where each (W_i,S_i) is an irreducible Coxeter system. Notice that a Coxeter element c of (W,S) can in this case be written as $c = (c_1,\ldots,c_r)$ where

c_i is a Coxeter element of (W_i, S_i). Irreducible finite Coxeter systems will be easier for us to work with, in part because they have been completely classified (see Table 1).

Proposition 5.1.1 *Let* $(W, S) = (W_1 \times \cdots \times W_r, S_1 \amalg \cdots \amalg S_r)$ *with* (W_i, S_i) *finite and irreducible. Then*

1. *$\Sigma(W, S)/\langle c \rangle$ is a pseudomanifold except in the following instance: $(W_i, S_i) = A_1$ for some i ($i = 1$ without loss of generality) and for all $i > 1$, (W_i, S_i) has an odd Coxeter number (hence from Table 1 we must have $(W_i, S_i) = A_{2k}$ or $I_2(2l + 1)$ for $i > 1$).*

2. *$\Sigma(W, S)/\langle c \rangle$ is an orientable pseudomanifold if and only if $\#S$ is even.*

Proof:

1. By Proposition 2.4.2, $\Sigma(W, S)/\langle c \rangle$ is a pseudomanifold except when $t \in \langle c \rangle$ for some reflection t, so assume $t = c^k$ for some k. Decompose $c = (c_1, \cdots, c_r)$, with c_i a Coxeter element of (W_i, S_i), so $t^k = (c_1^k, \ldots, c_r^k)$. We will use the following fact (see [Bo], Chapitre V, Section 6): if (W, S) is irreducible and finite and $\#S > 1$, then there exists a 2-plane in V on which c acts as a rotation through an angle of $\frac{2\pi}{h}$. Since t is a reflection, its multiset of eigenvalues is $\{-1, +1, +1, \ldots, +1\}$. By the above fact, for each i with $\#S_i > 1$, either c_i^k will either have two non-unit eigenvalues, or else $c_i^k = 1$. Thus in order to match t's eigenvalues, we must have $\#S_i = 1$ for exactly one i (and we may choose $i = 1$ without loss of generality). So $(W_1, S_1) = A_1$, and we have $c = (s, c_2, \ldots, c_r)$ and $t = (s^k, c_2^k, \ldots, c_r^k)$. Again to match t's eigenvalues, we must have $c_2^k = \cdots c_r^k = 1$ and $s^k = s$. Thus k must

(W,S)	Coxeter Diagram	Coxeter Number	Exponents
A_n		$n+1$	$1, 2, 3, \ldots, n$
B_n		$2n$	$1, 3, 5, \ldots, 2n-1$
D_n		$2(n-1)$	$1, 3, 5, \ldots, 2n-3, n-1$
E_6		10	$1, 4, 5, 7, 8, 9$
E_7		18	$1, 5, 7, 9, 11, 13, 17$
E_8		30	$1, 7, 11, 13, 17, 19, 23, 29$
F_4		12	$1, 5, 7, 11$
H_3		10	$1, 5, 9$
H_4		30	$1, 11, 19, 29$
$I_2(m)$		m	$1, m-1$

Table 5.1: Classification of irreducible finite Coxeter systems

be odd, and $c_2, \ldots, c_r$ must have odd orders. But this is exactly the instance described in 1.

2. By Proposition 2.4.2, $\Sigma(W, S)/\langle c \rangle$ is an orientable pseudomanifold if and only if $\operatorname{sgn}(g) = 1$ for all $g \in \langle c \rangle$. But $\operatorname{sgn}(c) = \operatorname{sgn}(s_1 \cdots s_m) = (-1)^{\#S}$, so this occurs exactly when $\#S$ is even.∎

Corollary 5.1.2 *If $\#S$ is even then*

$$\beta_J(\Sigma(W, S)/\langle c \rangle) = \beta_{S-J}(\Sigma(W, S)/\langle c \rangle) \ \forall J \subseteq S.$$

Proof: Apply Proposition 2.4.4 along with the previous proposition. ∎

One might ask if there are any other relations that hold among β_J's for $\Sigma(W, S)/\langle c \rangle$. One way in which they can arise is from symmetries of the Coxeter diagram.

Definition: Let (W, S) be a Coxeter system. The *Coxeter diagram* of (W, S) is the graph with vertex set S and having an edge labelled m_{ij} between node s_i and node s_j if m_{ij} is the order of $s_i s_j$ in W. When drawing the diagram (as in Table 1), it is conventional to omit the edges labelled 2, and omit the labels on edges labelled 3. A *diagram automorphism* of (W, S) is a bijection $\phi : S \to S$ such that for all i, j, $s_i s_j$ and $\phi(s_i)\phi(s_j)$ have the same order in W (and hence ϕ is a symmetry of the Coxeter diagram as a graph with lablelled edges). Because the pairwise order relations $(s_i s_j)^{m_{ij}} = 1$ form a presentation of W as a group ([Bro], Chapter II, Section 4), a diagram automorphism ϕ induces a well-defined group automorphism $\tilde{\phi} : W \to W$.

Proposition 5.1.3 *Let ϕ be a diagram automorphism of the finite Coxeter system (W, S). Then*

$$\beta_J(\Sigma(W,S)/\langle c \rangle\,) = \beta_{\phi(J)}(\Sigma(W,S)/\langle c \rangle\,) \ \forall J \subseteq S.$$

Proof: We will show that

$$\alpha_J(\Sigma(W,S)/\langle c \rangle\,) = \alpha_{\phi(J)}(\Sigma(W,S)/\langle c \rangle\,) \ \forall J \subseteq S$$

and the result then follows. By the fact mentioned after the first definition of this section, since $\tilde{\phi}(c)$ is another Coxeter element, we have $\tilde{\phi}(c) = u^{-1}cu$ for some $u \in W$. Define a map

$$\psi : W \to \{\text{double cosets } \langle c \rangle w W_{\phi(J)}\}$$

by

$$\psi(w) = \tilde{\phi}(\tilde{\phi}^{-1}(u)\langle c \rangle w W_J) = u \cdot u^{-1}\langle c \rangle u\phi(w)W_{\phi(J)} = \langle c \rangle u\tilde{\phi}(w)W_{\phi(J)}.$$

The first expression above for ϕ shows that it actually induces a well-defined map

$$\tilde{\psi} : \{\text{double cosets } \langle c \rangle w W_J\} \to \{\text{double cosets } \langle c \rangle w W_{\phi(J)}\}.$$

This first expression also shows that $\tilde{\psi}$ is bijective (since $\tilde{\phi}$ is an automorphism), and so we are done.∎

When $\#S$ is even, Corollary 5.1.2 tells us about a duality between β_J and β_{S-J} for $\Sigma(W,S)/\langle c \rangle$. Is there anything we can say when $\#S$ is odd? In many instances, we still have a form of weaker "local duality".

Definition: Given a subgroup G of W and $s \in S$, we will say G is *s-dual* if

$$\text{sgn}|_{w^{-1}Gw \cap W_{S-s}} = 1 \ \forall w \in W.$$

Proposition 5.1.4 (s-local duality) *If G is s-dual for some $s \in S$, then (abbreviating $\beta_J(\Sigma(W,S)/\langle c \rangle)$) by β_J) for all $J \subseteq S - s$, we have that*

$$\beta_J + \beta_{J+s} = \beta_{S-J} + \beta_{S-J-s}.$$

Proof:

$$
\begin{aligned}
\beta_J + \beta_{J+s} &= \sum_{K \subseteq J} (-1)^{\#(J-K)} \alpha_K + \sum_{K \subseteq J+s} (-1)^{\#(J+s-K)} \alpha_K \\
&= \sum_{s \in K \subseteq J+s} (-1)^{\#(J+s-K)} \alpha_K \\
&= \sum_{L \subseteq J} (-1)^{\#(J-L)} \alpha_{L+s}
\end{aligned}
$$

Now by definition

$$
\begin{aligned}
\alpha_{L+s} &= \#\{\text{double cosets } GwW_{S-L-s} \subseteq W\} \\
&= \sum_{\substack{double\ cosets \\ Gw_i W_{S-s} \subseteq W}} \#\{\text{double cosets } GwW_{S-L-s} \subseteq Gw_i W_{S-s}\}.
\end{aligned}
$$

We interject here a group-theory lemma that will help us to re-interpret this sum.

Lemma 5.1.5 *Let W be a finite group with subgroups G, H, I and $H \subseteq I$. Given $z \in W$, we have*

$$\#\{(z^{-1}Gz \cap I)xH \subseteq I : x \in I\} = \#\{GyH \subseteq GzI : y \in W\}.$$

Proof of lemma: Define a set map

$$\psi : I \rightarrow \{GyH \subseteq GzI : y \in W\}$$

by $\psi(x) = GzxH$. Clearly $\psi(x)$ depends only on the double coset $(z^{-1}Gz \cap I)xH$, and hence ψ induces a well-defined map $\tilde{\psi}$ between the two sets in the statement of this lemma.

$\tilde{\psi}$ *is surjective*: Given $GyH \subseteq GzI$, we have $y \in GzI$ and hence we can write $y = gzx$ for some $x \in I$. Then $\psi(x) = GzxH = GyH$ as we want.

$\tilde{\psi}$ *is injective*: Assume $Gzx_1H = Gzx_2H$ for some $x_1, x_2 \in I$. Then $x_1 \in z^{-1}Gzx_2H$, so we can write $x_1 = \gamma x_2 h$ with $\gamma \in z^{-1}Gz, h \in H$. But then $\gamma = x_1 h^{-1} x_2^{-1} \in I$, so $x_1 \in (z^{-1}Gz \cap I)x_2H$ as we want.∎

Continuing the proof of Proposition 5.1.4, we apply this lemma with

$$W = W, G = G, H = W_{S-L-s}, I = W_{S-s}, \text{ and } z = w_i$$

to conclude that $\alpha_{L+s} =$

$$\sum_{\substack{double\ cosets \\ Gw_i W_{S-s} \subseteq W}} \#\{\text{double cosets } (w_i^{-1}Gw_i \cap W_{S-s})wW_{S-L-s} \subseteq W_{S-s} : w \in W_{S-s}\}$$

$$= \sum_{\substack{double\ cosets \\ Gw_i W_{S-s} \subseteq W}} \alpha_L(\Sigma(W_{S-s}, S-s)/w_i^{-1}Gw_i \cap W_{S-s}).$$

Therefore

$$\beta_J + \beta_{J+s}$$

$$= \sum_{L \subseteq J} (-1)^{\#(J-L)} \sum_{\substack{double\ cosets \\ Gw_i W_{S-s} \subseteq W}} \alpha_L(\Sigma(W_{S-s}, S-s)/w_i^{-1}Gw_i \cap W_{S-s}).$$

$$= \sum_{\substack{double\ cosets \\ Gw_i W_{S-s} \subseteq W}} \beta_J(\Sigma(W_{S-s}, S-s)/w_i^{-1}Gw_i \cap W_{S-s}).$$

$$= \sum_{\substack{double\ cosets \\ Gw_i W_{S-s} \subseteq W}} \beta_{S-J-s}(\Sigma(W_{S-s}, S-s)/w_i^{-1}Gw_i \cap W_{S-s}).$$

$$= \beta_{S-J-s} + \beta_{S-J}.$$

The second-to-last equality comes from the assumption that G is s-dual (and Proposition 2.4.4). The last equality follows from reversing all the previous steps.∎

Our next result asserts that among finite Coxeter systems and $s \in S$, the property of $\langle c \rangle$ being s-dual is the rule rather than the exception.

Proposition 5.1.6 *Let* $(W, S) = (W_1 \times \cdots \times W_r, S_1 \amalg \cdots \amalg S_r)$ *with* (W_i, S_i) *finite and irreducible. Let* $s \in S_i$. *Then* $\langle c \rangle$ *is* s-dual *except when* $\#S$ *is odd and one of the following holds:*

1. $(W_i, S_i) = A_{n-1} \cong (S_n, \{(12), (23), \ldots, (n-1 \; n)\})$ *and* $s = (j \; j+1)$ *where* j *has odd order (additively) in* $\mathbf{Z}_n$

2. $(W_i, S_i) = I_2(m)$ *with* m *odd*

3. $(W_i, S_i) = E_6$ *and* s *is either one of the simple reflections which are farthest from each other in the Coxeter diagram of* E_6 *(see Table 1).*

Proof: Suppose $x \in w^{-1}\langle c \rangle w \cap W_{S-s}$ for some $s \in S, w \in W$. Then $x = w^{-1}c^l w$ for some l, and $c^l \in wW_{S-s}w^{-1}$. Clearly if $\#S$ is even then $\operatorname{sgn}(x) = \operatorname{sgn}(c)^l = 1$, so $\langle c \rangle$ will always be s-dual. Thus we may assume that $\#S$ is odd. Write $c = (c_1, \ldots, c_r)$ with c_k a Coxeter element for (W_k, S_k), and write $w = (w_1, \ldots, w_r)$. From $c^l \in wW_{S-s}w^{-1}$, we conclude that $c_i^l \in w_i W_{S_i - s} w_i^{-1}$. This implies that c_i^l fixes some non-zero vector $v \in V$ (e.g. let $v = w_i(v')$ where v' is constructed to be orthogonal to α if $r_\alpha \in S_i - s$). Hence c_i must have some eigenvalue λ for which $\lambda^l = 1$, which means (W_i, S_i) has an exponent e for which $le \equiv 0 \mod h$. We now check cases using the data from Table 1.

Case 1: (W_i, S_i) has even Coxeter number h, and all odd exponents e_j (this condition holds for $B_n, D_{2m}, E_7, E_8, F_4, H_3, H_4, I_2(2m)$). In this case $le \equiv 0 \mod h$ implies that l is even and hence $\operatorname{sgn}(x) = \operatorname{sgn}(c)^l = 1$, so $\langle c \rangle$ is s-dual.

Case 2: $(W_i, S_i) = D_{2m+1}$. If e is an odd exponent, then as in Case 1 we have $\text{sgn}(x) = 1$ so $\langle c \rangle$ is s-dual. If e is an even exponent, then $e = m$ and $h = 2m$. Thus $le \equiv 0 \bmod h$ implies l is even, and as in Case 1, $\langle c \rangle$ is s-dual.

Case 3: $(W_i, S_i) = A_{n-1} \cong (S_n, \{(12), (23), \ldots, (n-1\ n)\})$. Here we have $c = (12 \cdots n)$ and

$$c^l =$$

$$\left(1\ l+1 \cdots (\frac{n}{l} - 1)l + 1\right)\left(2\ l+2 \cdots (\frac{n}{l} - 1)l + 2\right) \cdots (n\ 2n \cdots n) =$$

$$w\left(1\ 2\ 3 \cdots \frac{n}{l}\right)\left(\frac{n}{l} + 1\ \frac{n}{l} + 2 \cdots 2\frac{n}{l}\right) \cdots \left((l-1)\frac{n}{l} + 1(l-1)\frac{n}{l} + 2 \cdots n\right) w^{-1}$$

for some $w \in W$. Thus $c^l \in w^{-1} W_{S-s} w$ for some w exactly when $s = (j\ j+1)$ for some j with order l (additively) in $\mathbf{Z}/n$. As before, $\langle c \rangle$ will be s-dual except if l is odd (since $\text{sgn}(x) = \text{sgn}(c)^l$), which is exactly the first exceptional case in the proposition.

Case 4: $(W_i, S_i) = I_2(m)$ with m odd. $I_2(m)$ is the *dihedral group of order* $2m$ with generators $\{s_1, s_2\}$ and relation $(s_1 s_2)^m = 1$. Since $c_i = s_1 s_2$, in this case we can have $c_i^l \in w_i W_{S-s} w_i^{-1}$ if and only if $l = m$. Hence $\text{sgn}(x) = \text{sgn}(c)^m = -1$ since m and $\#S$ are odd. Thus neither of $\langle c \rangle$ is neither s_1- nor s_2-dual. This is exactly the second exceptional case.

Case 5 $(W_i, S_i) = E_6$. Here (W_i, S_i) has exponents $1, 4, 5, 7, 8, 11$ and Coxeter number $h = 12$. As in Case 1, if e is an odd exponent then $\langle c \rangle$ is s-dual. Since the even exponents $e = 4, 8$ both satisfy $3e \equiv 0 \bmod 12$, we need only check for which $s \in S_i$ do we have $c^3 \in w_i W_{S-s} w_i^{-1}$ for some $w_i \in W_i$. Using the results of [Ca], one can show that this occurs exactly when s is as described in the third exceptional case of the proposition, but we omit the details.■

Example: Let $(W, S) = (S_4, \{(12), (23), (34)\})$. By brute force, one may calculate the table below for $\Sigma(W, S)/\langle c \rangle$.

J	α_J	β_J
$\emptyset$	1	1
(12)	1	0
(23)	2	1
(34)	1	0
$(12),(23)$	3	1
$(12),(34)$	3	2
$(23),(34)$	3	1
$(12),(23),(34)$	6	0

Alternatively, one could use the relations given by Propositions 5.1.3 and 5.1.4 to reduce the work. There is a single non-trivial diagram automorphism $\phi : (12) \mapsto (34), (23) \mapsto (23)$, so Proposition 5.1.3 tells us that

$$\beta_{(12)} = \beta_{(34)}$$

$$\beta_{(12),(23)} = \beta_{(23),(34)}.$$

By our last proposition, we see that for all $s \in S$, $\langle c \rangle$ is s-dual, and hence we have

$$
\begin{aligned}
\beta_\emptyset + \beta_{(12)} &= \beta_{(23),(24)} + \beta_{(12),(23),(34)} \\
\beta_{(23)} + \beta_{(12),(23)} &= \beta_{(34)} + \beta_{(12),(34)} \\
\beta_\emptyset + \beta_{(23)} &= \beta_{(12),(34)} + \beta_{(12),(23),(34)} \\
\beta_{(12)} + \beta_{(12),(23)} &= \beta_{(34)} + \beta_{(23),(34)} \\
\beta_\emptyset + \beta_{(34)} &= \beta_{(12),(23)} + \beta_{(12),(23),(34)} \\
\beta_{(12)} + \beta_{(12),(34)} &= \beta_{(23)} + \beta_{(23),(34)}.
\end{aligned}
$$

This gives a total of 8 linear relations, however one can check that the last 4 are linear combinations of the first 4. Since there are 8 β_J's, we only need to calculate 4 of them in order to fill in the rest using these relations. Since we always have $\beta_\emptyset = 1$, and in this case $\beta_S = 0$ (via Proposition 2.4.1), we only need to calculate 2 further β_J's by brute force, e.g. $\beta_{(12)}, \beta_{(23)}$.

5.2   Primitivity

It would be very desirable to have a partitioning or shelling of $\Sigma(W, S)/\langle c \rangle$ like the ones in Chapter 4 for $\Sigma(W^r, rS)/\Delta^r(W')$ and $\Sigma(W^r, rS)/\Delta^r(E')$. This however seems to be a difficult problem. In fact the next example shows that $\Sigma(W, S)/\langle c \rangle$ is not in general shellable, but even a general partitioning of $\Sigma(W, S)/\langle c \rangle$ has eluded us.

Example: Let $(W, S) = (S_n , \{(12), (23), \ldots, (n-1\ n)\})$ with n *prime.* Then $c = (12 \cdots n)$, and it is not hard to see (using the description of $\Sigma(W, S)/\langle c \rangle$ as the barycentric subdivision of a simplex having vertices $\{1, 2, \ldots, n\}$) that $\langle c \rangle$ gives a free $\mathbf{Z}_n$-action on the sphere. By the same reasoning as in the example after Theorem 4.3.5, we conclude that the quotient $\Sigma(W, S)/\langle c \rangle$ is not shellable.

Even though $\Sigma(W, S)/\langle c \rangle$ has not been partitioned, there is a large "chunk" of $\Sigma(W, S)/\langle c \rangle$ (the primitive part) which is in some instances more tractable.

Definition: We will say a face wW_J of $\Sigma(W, S)/\langle c \rangle$ is *primitive (with respect* *to c)* if $c^i w W_J \neq w W_J$ unless $c^i = 1$, or in other words, $\langle c \rangle \cap w W_J w^{-1} = 1$. We will say a face $\langle c \rangle w W_J$ of $\Sigma(W, S)/\langle c \rangle$ is *primitive* if $w W_J$ is primitive (this clearly only depends on $\langle c \rangle w W_J$). We will say a vector $f \in V$ is *primitive* if $c^i(f) \neq f$ unless $c^i = 1$, and an orbit $\langle c \rangle f$ of vectors is *primitive* if f is primitive.

One can easily check that f or $\langle c \rangle f$ is primitive if and only if $F(f)$ or $F(\langle c \rangle f) = \langle c \rangle F(f)$ is primitive, respectively. We will let $\Sigma(W, S)_{prim}$ denote the subposet of $\Sigma(W, S)$ consisting of all primitive faces $w W_J$, and similarly for $\Sigma(W, S)/\langle c \rangle_{prim}$.

Our next proposition gives (in some cases) a fundamental domain for the action of $\langle c \rangle$ on the primitive vectors of V.

Proposition 5.2.1 *Let (W, S) be one of the infinite families A_n, B_n, D_n, $I_2(m)$ of finite irreducible Coxeter systems. Let*

$$A = \{(b, f) \in W \times V : b \text{ is a conjugate } u^{-1} c u \text{ of } c, \text{ and } b(f) \in \mathcal{A}(b\Phi^+) \}.$$

Then the map $\phi : A \to \{orbits \ \langle c \rangle f' : f' \in V\}$ given by $\phi(b, f) = \langle c \rangle u(f)$ is well-defined, and a bijection onto $\{primitive \ \langle c \rangle f' : f' \in V\}$.

Proof: To show ϕ is well-defined, we need to see that if $b = u^{-1} c u = v^{-1} c v$ for some $u, v \in W$, then $\langle c \rangle u(f) = \langle c \rangle v(f)$. But $u^{-1} c u = v^{-1} c v$ implies $v u^{-1}$ commutes with c, and it is known ([Ca], Proposition 30) that if (W, S) is irreducible then the centralizer of c in W is $\langle c \rangle$. So $v u^{-1} \in \langle c \rangle$ and $\langle c \rangle u(f) = \langle c \rangle v(f)$.

The fact that ϕ is a bijection onto the primitive orbits for $(W, S) = A_n$ is a special case of result of Gessel (mentioned in [Ge1] and described in [DeWa], Section 3). We will give Gessel's construction of ϕ^{-1}, leaving it to the reader to verify the rest of the details (namely that $\phi(b, f)$ is always primitive, and that $\phi^{-1}(\langle c \rangle f') = (b, f)$ always satisfies $b(f) \in \mathcal{A}(b\Phi^+)$). We will then mimic this construction of ϕ^{-1} for $(W, S) = B_n, D_n$. The case of $(W, S) = I_2(m)$ is easy to check using the description of $I_2(m)$ as the dihedral group acting on $\mathbf{R}^2$, with c acting as rotation through $\frac{2\pi}{n}$.

Gessel's construction of ϕ^{-1} for $A_{n-1} \cong S_n$ goes as follows. In this case, $c = (12 \cdots n)$. Given $\langle c \rangle f'$ primitive with $(f_1', \ldots, f_n') \in V$, let w_i for $i = 1, \ldots, n$ be the word of length n defined by

$$w_i = f_i' c^{-1}(f')_i c^{-2}(f')_i \cdots c^{-n+1}(f')_i$$

(i.e. w_i is the sequence of numbers that pass through the ith coordinate as one repeatedly applies c^{-1} to f', or in other words, w_i is the word gotten by reading f' starting from the ith coordinate and moving to the right with a wraparound from f_n' to f_1'). Rank these words $\{w_i\}_{i=1,\ldots n}$ in lexicographic order from largest to smallest (primitivity of $\langle c \rangle f'$ assures that no two of them are equal). Let r_i be the rank of w_i, e.g. if w_1 is third largest lexicographically, then $r_1 = 3$. Then $u^{-1} = \begin{pmatrix} 1 & \cdots & n \\ r_1 & & r_n \end{pmatrix}$ and $\phi^{-1}(\langle c \rangle f') = (u^{-1}cu, u^{-1}(f'))$. For example, let $n = 8$ and $f' = (1, 2, 2, 1, 4, 3, 2, 4)$. We then have

$$w_1 = 12214324, w_2 = 22143241, w_3 = 21432412, \text{ etc.}$$

and the ranking is

$$w_5 \geq w_8 \geq w_6 \geq w_7 \geq w_2 \geq w_3 \geq w_4 \geq w_1.$$

So $u^{-1} = \begin{pmatrix} 12345678 \\ 85671342 \end{pmatrix}$ and

$$\phi(\langle c \rangle f') = (b, f) = (u^{-1}cu, u^{-1}(f'))$$

$$= \left((85671342) = \begin{pmatrix} 12354768 \\ 38426715 \end{pmatrix}, (4,4,3,2,2,2,1,1) \right).$$

Notice in this example that f satisfies $f_1 \geq \ldots \geq f_n$, and $f_i > f_{i+1}$ whenever $(i \ i+1) \in D(b)$, which are the same conditions as $b(f) \in \mathcal{A}(b\Phi^+)$.

We now vary this construction for B_n. Here W is the set of all permutations and *sign changes* on the coordinates in $V = \mathbf{R}^n$, with simple reflections

$$S = \left\{ (12), (23), \ldots, (n\ n-1), \begin{pmatrix} n \\ -n \end{pmatrix} \right\}$$

and

$$c = (12)(23) \cdots (n\ n-1) \begin{pmatrix} n \\ -n \end{pmatrix} = \begin{pmatrix} 1\ 2 & & n-1 & n \\ 2\ 3 & \cdots & n & -1 \end{pmatrix}.$$

Given $\langle c \rangle f'$ primitive with $(f'_1, \ldots, f'_n) \in V$, again we let w_i for $i = 1, \ldots, n$ be the word of length n defined by

$$w_i = f'_i c^{-1}(f')_i c^{-2}(f')_i \cdots c^{-n+1}(f')_i$$

(i.e. w_i is the word gotten by reading f' starting from the ith coordinate and moving to the right with a wraparound and *persistent sign change* after f'_n). Whenever the first non-zero coordinate in w_i is negative, we negate all of w_i to make it positive. Rank these words $\{w_i\}_{i=1,\ldots,n}$ in lexicographic order from largest to smallest (as before, primitivity of $\langle c \rangle f'$ assures that no two of them are equal), and let r_i be the rank of w_i. Next we put signs on the positive numbers r_i to get integers r'_i as follows. If $f'_i \neq 0$, then r'_i has the same sign as f'_i. If $f'_i = 0$, then let r'_i have the same sign as r'_{i+1} if $i < n$, and opposite sign as r'_{i+1} if $i = n$. We then let $u^{-1} = \begin{pmatrix} 1 & \cdots & n \\ r'_1 & & r'_n \end{pmatrix}$ and $\phi^{-1}(\langle c \rangle f') = (u^{-1}cu, u^{-1}(f'))$. For example, let $n = 5$ and $f' = (+1, 0, -1, 0, +2)$. Before negations, we have

$$w_1 = +10 - 10 + 2, w_2 = 0 - 10 + 2 - 1, w_3 = -10 + 2 - 10,$$

$$w_4 = 0 + 2 - 10 + 1, w_5 = +2 - 10 + 10$$

and we must negate w_2, w_3 to get $w_2 = 0 + 10 - 2 + 1, w_3 = +10 - 2 + 10$. The ranking is

$$w_5 \geq w_1 \geq w_3 \geq w_4 \geq w_2,$$

so $(r_1, \ldots, r_5) = (2, 5, 3, 4, 1)$. Then $(r_1', \ldots, r_5') = (+2, -5, -3, +4, +1)$, so
$u^{-1} = \begin{pmatrix} 1 & 2 & 3 & 4 & 5 \\ +2 & -5 & -3 & +4 & +1 \end{pmatrix}$ and

$$\phi(\langle c \rangle f') = (b, f) = (u^{-1}cu, u^{-1}(f'))$$

$$= \left(\begin{pmatrix} 1 & 2 & 3 & 4 & 5 \\ -2 & +5 & -4 & +1 & +3 \end{pmatrix}, (+2, +1, +1, 0, 0) \right).$$

We now vary this construction for D_n. Here W is the set of all permutations and an *even* number of sign changes on the coordinates in $V = \mathbf{R}^n$, with simple reflections

$$S = \{(12), (23), \ldots, (n\ n - 1), \begin{pmatrix} n - 1 & n \\ -n & -(n - 1) \end{pmatrix} \}$$

and

$$c = (12)(23) \cdots (n\ n - 1) \begin{pmatrix} n - 1 & n \\ -n & -(n - 1) \end{pmatrix} = \begin{pmatrix} 1 & 2 & \cdots & n - 2 & n - 1 & n \\ 2 & 3 & \cdots & n - 1 & -1 & -n \end{pmatrix}.$$

Given $\langle c \rangle f'$ primitive with $(f_1', \ldots, f_n') \in V$, we define the words w_i and their rankings r_i as we did in the B_n construction. The difference lies in the signs we put on r_i to get the integers r_i'. If $r_i \neq n$ and $f_i' \neq 0$, then r_i' has the same sign as f_i'. If $r_i' \neq n$ and $f_i' = 0$, then there are three case depending on whether $i < n - 1, i = n - 1$, or $i = n$. The third case cannot occur, since then $w_n = 000 \ldots$, and hence $r_n = n$. In the first case, let r_i' have the same sign as r_{i+1}', and in the second case, let them have opposite signs. This only leaves r_i' undetermined when $r_i = n$, and we let

$r'_i = \pm n$ so as to make the total number of negative r'_i even. We then let $u^{-1} = \left(\begin{smallmatrix} 1 & \cdots & n \\ r'_1 & & r'_n \end{smallmatrix} \right)$ and $\phi^{-1}(\langle c \rangle f') = (u^{-1}cu, u^{-1}(f'))$. For example, let $n = 6$ and $f' = (-1, 0, 0, +2, +1, 0)$. Then

$$w_1 = +100 - 2 - 1, w_2 = 00 + 2 + 1 + 1, w_3 = 0 + 2 + 1 + 10,$$

$$w_4 = +2 + 1 + 100, w_5 = +1 + 100 - 2, w_6 = 000000.$$

The ranking is

$$w_4 \geq w_5 \geq w_1 \geq w_3 \geq w_2 \geq w_6,$$

so $(r_1, \ldots, r_6) = (3, 5, 4, 1, 2, 6)$. Then

$$(r'_1, \ldots, r'_6) = (-3, +5, +4, +1, +2, -6),$$

so $u^{-1} = \left(\begin{smallmatrix} 1 & 2 & 3 & 4 & 5 & 6 \\ -3 & +5 & +4 & +1 & +2 & -6 \end{smallmatrix} \right)$ and

$$\phi(\langle c \rangle f') = (b, f) = (u^{-1}cu, u^{-1}(f'))$$

$$= \left(\left(\begin{matrix} 1 & 2 & 3 & 4 & 5 & 6 \\ +2 & +3 & -5 & +1 & +4 & -6 \end{matrix} \right), (+2, +1, +1, 0, 0, 0) \right).$$

The previous proposition suggests the following conjecture:

Conjecture 5.2.2 *The map ϕ of the prevous proposition is a bijection as stated for all irreducible Coxeter systems (i.e. it holds for the exceptional groups $E_6, E_7, E_8, F_4, H_3, H_4$ as well), and with a more unified proof.*

Theorem 5.2.3 *For $(W, S) = A_n, B_n, D_n$ or $I_2(m)$, we have*

$$\Sigma(W, S)/\langle c \rangle_{prim} = \coprod_{\substack{cosets\ \langle c \rangle u \subseteq W}} [\langle c \rangle u W_{S - D(u^{-1}cu)}, \langle c \rangle u W_\emptyset].$$

Proof: According to the previous theorem, if we let $V/\langle c \rangle_{prim}$ be the set of all primitive orbits $\langle c \rangle f'$ in V, then

$$V/\langle c \rangle_{prim} = \{ \langle c \rangle u(f) : b \text{ is a conjugate } u^{-1}cu \text{ of } c, \text{ and } b(f) \in \mathcal{A}(b\Phi^+) \}$$

$$= \{ \langle c \rangle u(f) : \langle c \rangle u \subseteq W, u^{-1}cu(f) \in \mathcal{A}(u^{-1}cu\Phi^+) \}.$$

Applying the map $\langle c \rangle f' \mapsto F(\langle c \rangle f')$ to both sides, and using Lemma 3.2.1 gives the result.∎

Note that $\Sigma(W,S)_{prim}$ and $\Sigma(W,S)/\langle c \rangle_{prim}$ are only subposets of the simplicial posets $\Sigma(W,S)$ and $\Sigma(W,S)/\langle c \rangle$ respectively, and not simplicial posets themselves. Nevertheless we can still define α_J for both as usual, and then let $\beta_J = \sum_{K \subseteq J} (-1)^{\#(J-K)} \alpha_K$.

Corollary 5.2.4 *For $(W,S) = A_n, B_n, D_n$ or $I_2(m)$, we have*

$$\beta_J(\Sigma(W,S)/\langle c \rangle_{prim}) = \#\{b \in W : D(b) = J, b \text{ conjugate to } c\}.$$

Proof: same as proof of Proposition 3.2.3.∎

Remark: Note that the Coxeter element c for (W^r, rS) is the same as the diagonal embedding $\Delta^r(c_1) = (c_1, c_1, \ldots, c_1)$ of the Coxter element c_1 for (W,S). Thus using our standard multipartite techniques from Sections 3.4 and 4.1, we can soup up the proof of Proposition 5.2.1 (replace all inequalities $\geq$ by $\geq_{\mathcal{L}}$) and prove that the same proposition holds if we replace V by V^r and $\mathcal{A}(\cdot)$ by $\mathcal{A}_r(\cdot)$. From this we can then deduce the following multipartite analogues of Theorem 5.2.3 and Corollary 5.2.4:

$$\Sigma(W^r, rS)/\langle c \rangle_{prim} =$$

$$\coprod_{\substack{cosets \ \langle c \rangle u \subseteq W}} \coprod_{\substack{(w_1,...,w_r) \in W^r \\ w_r \cdots w_1 = u^{-1} c_1 u}} \langle c \rangle \prod_{i=1}^{r} [w_r w_{r-1} \cdots w_i W_{S-D(w_i)}, w_r w_{r-1} \cdots w_i W_{\emptyset}]$$

$$\beta_{(J_1,...,J_r)}(\Sigma(W^r, rS)/\langle c \rangle_{prim}) =$$

$$\#\{(w_1,...,w_r) \in W^r : D(w_i) = J_i, w_r w_{r-1} \cdots w_1 \ conjugate \ to \ c_1\}.$$

Since we have a combinatorial interpretation for $\beta_J(\Sigma(W,S)/\langle c \rangle_{prim})$ in the instances above, we would like to know when to expect some kind of duality like $\beta_J = \beta_{S-J}$, as in the fine Dehn-Somerville equations. Since $\Sigma(W,S)/\langle c \rangle_{prim}$ is not even a simplicial poset, we cannot simply apply Proposition 2.4.4. Our strategy is to *filter* $\Sigma(W, S)/\langle c \rangle$ into pieces according to their *primitivity*, and then use Proposition 2.4.4 on the pieces.

Definition: Let (W, S) be a finite Coxeter system (not necessarily irreducible) with Coxeter number h, and a Coxeter element c. Given j dividing h, let

$$\Sigma_{\leq j} = \{wW_J \in \Sigma(W, S) : c^j wW_J = wW_J\}$$

$$\Sigma_{=j} = \Sigma_{\leq j} - \bigcup_{i|j} \Sigma_{\leq i}.$$

Note that $\Sigma_{=h} = \Sigma(W, S)_{prim}$.

Proposition 5.2.5

$$\beta_J(\Sigma(W, S)/\langle c \rangle_{prim}) = \frac{1}{h} \sum_{d|h} \mu(d)\beta_J(\Sigma_{\leq \frac{h}{d}})$$

for all $J \subseteq S$, where μ denotes the number-theoretic Möbius function ([HW], Section 16.3).

Proof: By linearity it suffices to prove the same result replacing β_J by α_J on both sides. Note that primitivity of wW_J is equivalent to the property that $\{c^i wW_J\}_{i=1,\dots,h}$ are all distinct. Hence we have

$$\alpha_J(\Sigma(W,S)/\langle c\rangle_{prim}) = \frac{1}{h}\alpha_J\Sigma(W,S)_{prim} = \frac{1}{h}\alpha_J(\Sigma_{=h}).$$

Since $\Sigma_{\leq j} = \coprod_{i|j}\Sigma_{=i}$ implies $\alpha_J(\Sigma_{\leq j}) = \sum_{i|j}\alpha_J(\Sigma_{=i})$, we can apply Möbius inversion ([HW], Section 16.4) to conclude that

$$\alpha_J(\Sigma_{=j}) = \sum_{i|j}\mu\left(\frac{j}{i}\right)\alpha_J(\Sigma_{\leq i}).$$

and hence

$$\alpha_J(\Sigma(W,S)/\langle c\rangle_{prim}) = \frac{1}{h}\sum_{i|h}\mu\left(\frac{h}{i}\right)\alpha_J(\Sigma_{\leq i}).$$

Replacing $\frac{h}{i}$ by d gives what we wanted.∎

Theorem 5.2.6 *In the following instances, we have*

$$\beta_J(\Sigma(W,S)/\langle c\rangle_{prim}) = \beta_{S-J}(\Sigma(W,S)/\langle c\rangle_{prim})$$

for all $J \subseteq S$:

1. *$(W,S) = A_{n-1}$ and $n \not\equiv 2 \mod 4$*

2. *$(W,S) = B_n$ and n even*

3. *$(W,S) = D_n$ and $n-1$ a power of 2*

4. *$(W,S) = I_2(m)$*

Proof: The instances above all share the following property: for all $d|h$, the subposet $\Sigma_{\leq \frac{h}{d}}$ is a balanced simplicial poset triangulating a sphere and having label set R for some subset R of S. In this situation, we have

$$
\begin{aligned}
\beta_J(\Sigma_{\leq \frac{h}{d}}) &= \sum_{K \subseteq J} (-1)^{\#(J-K)} \alpha_K(\Sigma_{\leq \frac{h}{d}}) \\
&= (-1)^{\#(J-R)} \sum_{K \subseteq J \cap R} (-1)^{\#(R-K)} \alpha_K(\Sigma_{\leq \frac{h}{d}}) \\
&= (-1)^{\#(J-R)} \beta_{J \cap R}(\Sigma_{\leq \frac{h}{d}})
\end{aligned}
$$

and hence by Proposition 2.4.4 applied to the sphere $\Sigma_{\leq \frac{h}{d}}$, we have

$$
\begin{aligned}
\beta_J(\Sigma_{\leq \frac{h}{d}}) &= (-1)^{\#(J-R)} \beta_{R-(J \cap R)}(\Sigma_{\leq \frac{h}{d}}) \\
&= (-1)^{\#(J-R)} \beta_{(S-J) \cap R}(\Sigma_{\leq \frac{h}{d}}) \\
&= (-1)^{\#(J-R)+\#(S-J-R)} \beta_{S-J}(\Sigma_{\leq \frac{h}{d}}) \\
&= (-1)^{\#(S-R)} \beta_{S-J}(\Sigma_{\leq \frac{h}{d}})
\end{aligned}
$$

We will apply this equality in each of the cases above, and using explicit descriptions for $\Sigma(W,S)$ in each case.

1. $(W,S) = A_{n-1}, n \not\equiv 2 \bmod 4$. Here $c = (12 \cdots n)$, $h = n$ and $c^{\frac{n}{d}}$ has $\frac{n}{d}$ cycles, each of size d. $\Sigma(W,S)$ is the barycentric subdivision of the boundary of an $(n-1)$-simplex, and thus has vertices corresponding to the subsets of $\{1,2,\ldots,n\}$. From this we see that $\Sigma_{\leq \frac{n}{d}}$ is the barycentric subdivision of the boundary of the $(\frac{n}{d}-1)$-simplex having vertices corresponding to all unions of orbits of $c^{\frac{n}{d}}$. Thus the label set $R \subseteq S$ for $\Sigma_{\leq \frac{n}{d}}$ has $\#R = \frac{n}{d} - 1$, and we have

$$
(-1)^{\#(S-R)} = (-1)^{(n-1)-(\frac{n}{d}-1)} = (-1)^{\frac{n}{d}(d-1)}.
$$

Thus

$$\beta_J(\Sigma(W,S)_{prim}) = \sum_{d|n} \mu(d)\beta_J(\Sigma_{\leq \frac{n}{d}})$$

$$= \sum_{d|n} \mu(d)(-1)^{\frac{n}{d}(d-1)}\beta_{S-J}(\Sigma_{\leq \frac{n}{d}}).$$

Hence our conclusion will follow when $\mu(d)(-1)^{\frac{n}{d}(d-1)} = \mu(d)$ for all d dividing n. One can check that this occurs exactly when $n \not\equiv 2 \bmod 4$.

2. $(W,S) = B_n$ with n even. Here $c = \begin{pmatrix} 1 & 2 & \cdots & n-1 & n \\ 2 & 3 & & n & -1 \end{pmatrix}, h = 2n,$ and $c^{\frac{2n}{d}}$ acts differently depending on the parity of d. If d is odd, then $c^{\frac{2n}{d}}$ has $\frac{n}{d}$ cycles, each of size d, and in which the total number sign changes in each cycle is even. For example, if $n = 6$ and $d = 3$ then

$$c^{\frac{2n}{d}} = c^4 = \begin{pmatrix} 1 & 3 & 5 \\ 5 & -1 & -3 \end{pmatrix}\begin{pmatrix} 2 & 4 & 6 \\ 6 & -2 & -4 \end{pmatrix}$$

If d is even, then $c^{\frac{2n}{d}}$ has an odd number of sign changes in each of its cycles. $\Sigma(W,S)$ is the barycentric subdivision of the boundary of an n-cube, whose vertices correspond to all signed subsets of $\{\pm 1, \ldots, \pm n\}$ (i.e. all subsets of the previous set that contain no pair i and $-i$). Thus when d is odd, we get a pair of opposite signed subsets for each cycle of $c^{\frac{2n}{d}}$, and unions of these over all the cycles give the vertices of $\Sigma_{\leq \frac{2n}{d}}$, which is the barycentric subdivision of the boundary of an $\frac{n}{d}$-cube. Thus if d is even, the label set R of $\Sigma_{\leq \frac{2n}{d}}$ has $\#R = \frac{n}{d}$, and $(-1)^{\#(S-R)} = (-1)^{\frac{n}{d}(d-1)}$. If d is even, $c^{\frac{2n}{d}}$ fixes only the empty signed subset, so $\#R = 0$ and $(-1)^{\#(S-R)} = (-1)^n$. Therefore,

$$\beta_J(\Sigma(W,S)_{prim}) = \sum_{\substack{d|2n \\ d \text{ odd}}} \mu(d)\beta_J(\Sigma_{\leq \frac{2n}{d}})$$

$$= \quad + \sum_{\substack{d|2n \\ d \ even}} \mu(d)\beta_J(\Sigma_{\leq \frac{n}{d}})$$

$$= \quad \sum_{\substack{d|2n \\ d \ odd}} \mu(d)(-1)^{\frac{n}{d}(d-1)}\beta_{S-J}(\Sigma_{\leq \frac{2n}{d}})$$

$$+ \sum_{\substack{d|2n \\ d \ even}} \mu(d)(-1)^n \beta_{S-J}(\Sigma_{\leq \frac{n}{d}})$$

Hence our conclusion will follow when $\mu(d)(-1)^{\frac{n}{d}(d-1)} = \mu(d)$ for all odd d dividing $2n$ and $\mu(d)(-1)^n = \mu(d)$ for all even d dividing $2n$. A bit of thought shows that this is true exactly when n is even.

3. $(W, S) = D_n$ with $n - 1$ a power of 2. Here $c = \begin{pmatrix} 1 & \cdots & n-2 & n-1 & n \\ 2 & & n-1 & -1 & -n \end{pmatrix}, h = 2(n - 1)$ and $d|h$ implies that d is a power of 2. One can check that $c^{\frac{2(n-1)}{d}}$ breaks up into the singleton cycle $\binom{n}{n}$, and all other cycles having an odd number of sign changes. $\Sigma(W, S)$ is the subdivsion of the boundary of the n-cube having vertices corrseponding to all signed subsets of $\{\pm 1, \ldots, \pm n\}$ *except* for those of cardinality $n - 1$. Hence $\Sigma_{\leq \frac{2(n-1)}{d}}$ is the 0-sphere with vertices corresponding to the signed sets $\{n\}, \{-n\}$. Thus $\#R = 1$, and $(-1)^{\#(S-R)} = (-1)^{n-1}$. Therefore,

$$\beta_J(\Sigma(W, S)_{prim}) \quad = \quad \sum_{d|2(n-1)} \mu(d)\beta_J(\Sigma_{\leq \frac{2(n-1)}{d}})$$

$$= \quad \sum_{d|2(n-1)} \mu(d)(-1)^{n-1}\beta_{S-J}(\Sigma_{\leq \frac{2(n-1)}{d}})$$

But $(-1)^{n-1} = 1$ since $n - 1$ is a power of 2, so the result follows.

4. $(W, S) = I_2 m$. Then c is the rotation through $\frac{2\pi}{m}$ acting on $\Sigma(W, S)$, which is the barycentric subdivision of a regular m-gon in the plane.

It is easy to compute directly that for $\Sigma(W,S)/\langle c\rangle_{prim}$ we have

$$\beta_\emptyset = \beta_S = 0, \ \beta_{s_1} = \beta_{s_2} = 1$$

where $S = \{s_1, s_2\}$.∎

Remark: Once can do a similar analysis for $\Sigma(W^r, rS)_{prim}$ and get that $\beta_{J_1,\ldots,J_r} = \beta_{S-J_1,\ldots,S-J_r}$ for all $J_i \subseteq S$ in the following instances:

1. $(W,S) = A_{n-1}$ and either r even or $n \not\equiv 2 \bmod 4$

2. $(W,S) = B_n$ and either r even or n is even

3. $(W,S) = D_n$ and $n-1$ a power of 2

4. $(W,S) = I_2(m)$.

Corollary 5.2.7 *In the instances listed in the above remark, for all $J_i \subseteq S$ we have*

$$\#\{(w_1,\ldots,w_r) \in W^r : D(w_i) = J_i, w_r w_{r-1}\cdots w_1 \text{ is conjugate to } c\} =$$

$$\#\{(w_1,\ldots,w_r) \in W^r : D(w_i) = S - J_i, w_r w_{r-1}\cdots w_1 \text{ is conjugate to } c\}.$$

Proof: Combine Theorem 5.2.6 (and its succeeding remark) with Corollary 5.2.4 (and its succeeding remark).∎

Remark: As in remarks after Corollary 4.1.5, Gessel (personal communication) has shown how to prove an even stronger result in the case of $W = S_n$

using the theory of symmetric functions. Presumably an analogous technique might work for the cases of B_n and D_n.

Example: Let $(W, S) = A_2 = (S_3, \{(12), (23)\}), r = 2$. We can compile the following list of pairs of permutations (w_1, w_2) such that $w_2 w_1$ is conjugate to $c = (123)$:

$$(123, 23 \cdot 1) \qquad (3 \cdot 2 \cdot 1, 2 \cdot 13)$$
$$(13 \cdot 2, 2 \cdot 13) \qquad (2 \cdot 13, 13 \cdot 2)$$
$$(2 \cdot 13, 3 \cdot 2 \cdot 1) \quad (23 \cdot 1, 123)$$
$$(13 \cdot 2, 3 \cdot 2 \cdot 1) \quad (3 \cdot 12, 123)$$
$$(23 \cdot 1, 23 \cdot 1) \qquad (3 \cdot 12, 3 \cdot 12)$$
$$(123, 3 \cdot 12) \qquad (3 \cdot 2 \cdot 1, 13 \cdot 2)$$

Note that they have been listed so that pairs within the same row have complementary descent sets, illustrating an instance of the previous corollary.

Chapter 6

Acknowledgments

I would like to thank my thesis advisor Richard Stanley for numerous helpful comments and suggestions, and Ira Gessel for suggesting some of the probelms that originally got me started in this area. I am greatly indebted to Matthew Dyer for teaching me about reflection subgroups of Coxeter systems, and for all his patient technical help in that area. I also thank Ian Grojnowski for his Coxeter group expertise.

Appendix A

Technical results on reflection subgroups

Here we collect together most of the technical tools we have used in Chapters 3 and 4 concerning reflection subgroups. I am greatly indebted to Matthew Dyer for the proofs of all of these results. Many of these may be paraphrased as saying that "reflection subgroups behave almost as nicely as standard parabolic subgroups".

For the remainder of this appendix, let (W, S) be a Coxeter system (not necessarily finite) realized by the positive root system (Φ^+, Π) on an $\mathbf{R}$-vector space V, and let T be the reflections of W i.e.

$$T = \bigcup_{w \in W, s \in S} wsw^{-1} = \{r_\alpha : \alpha \in \Phi^+\}.$$

Let W' be a reflection subgroup of W, i.e. $W' = \langle W' \cap T \rangle$. Recall

$$I(w) = \{t \in T : l(tw) < l(w)\} = \{r_\alpha : \alpha \in \Phi^+ \cap w^{-1}(-\Phi^+)\}.$$

Definition: The *canonical generators* S' of W' are defined by

$$S' = \{t \in T : I(t) \cap W' = \{t\}\}.$$

Let $\Phi_{W'}^+ = \{\alpha \in \Phi^+ : r_\alpha \in W'\}$, $\Pi_{W'} = \{\alpha \in \Phi^+ : r_\alpha \in S'\}$, and let $V_{W'}$ be the **R**-span of $\Phi_{W'}^+$.

The next theorem justifies the notation just defined.

Theorem A.0.8 ([Dy], Theorem 3.3) (W', S') *is a Coxeter system realized by the positive root system* $(\Phi_{W'}^+, \Pi_{W'})$ *on* $V_{W'}$ *whose length function*

$$l'(w) = min\{r : w = s_1' s_2' \cdots s_r', s_i' \in S'\}$$

is given by $l'(w) = \#(I(w) \cap W')$.■

Definition: Let $D_{W'} = \{w \in W : I(w) \cap W' = \emptyset\}$ (this is the same as $\mathcal{L}(P(W'))$ from Chapters 3 and 4).

Proposition A.0.9 ([Dy], Corollary 3.4)

1. *Every* $w \in W$ *can be factored uniquely in the form* $w = xy$ *where* $x \in D_{W'}, y \in W'$.

2. *If* $y \in D_{W'}$, *then the map* $x \mapsto xy$ *from* W' *to* $W'y$ *is an isomorphism of Bruhat order* $\leq_B$. *Hence* y *is the unique element of minimal length in* $W'y$, *and the least element of* $W'y$ *in Bruhat order.*■

The next lemma will be used frequently in the proofs in this appendix.

Lemma A.0.10 ([Dy], Definition 2.1)

1. $I(xy) = I(x) + xI(y)x^{-1}$, *where + denotes the operation of symmetric difference of sets (i.e. $A + B = (A - B) \cup (B - A)$).*

2. *If $x \in D_{W_J}^{-1}$ and $y \in W_J$ for some $J \subseteq S$, then*

$$I(xy) = I(x) \amalg xI(y)x^{-1}. \blacksquare$$

Proposition A.0.11 (Dyer) *Fix $J \subseteq S$.*

1. *Every $w \in W$ can be factored uniquely in the form $w = xyz$ where $x \in W', y \in D_{W'} \cap D_{W_J}^{-1}, z \in W_J \cap D_{y^{-1}W'y}$.*

2. *In the above factorization, we have $xy \in D_{W'}$, and $l(xyz) > l(x)$ unless $x = z = 1$. Thus if $y \in D_{W'} \cap D_{W_J}^{-1}$, then y is the unique element of minimal length in $W'yW_J$, and the least element of $W'yW_J$ in Bruhat order.*

Proof: To prove 1, given $w, J,$ and W' we must show that there exists a unique such factorization.

Existence: Let y be an element of $W'yW_J$ of minimal length. Clearly then $y \in D_{W'} \cap D_{W_J}^{-1}$ and $w = x'yz'$ for some $x' \in W', z' \in W_J$. Now decompose $z' = z''z$ with $z'' \in y^{-1}W'y$ and $z \in D_{y^{-1}W'y \cap W_J}$ using Proposition A.0.9. Then

$$w = x'yz' = x'yz''z = x'y \cdot y^{-1}x''yzz = x'x''yz = xyz$$

where $x'' \in W'$.

Rather than showing uniqueness, let us first show 2. We have

$$
\begin{aligned}
I(yz) \cap W' &= I(y) \cap W' + yI(z)y^{-1} \cap W' \\
&= \emptyset + y(I(z) \cap y^{-1}W'y)y^{-1}
\end{aligned}
$$

$$= y(I(z) \cap y^{-1}W'y \cap W_J)y^{-1}$$

$$= \emptyset$$

The first equality comes from Lemma A.0.10, and the third from the fact that $z \in W_J$. Thus $xy \in D_{W'}$, and hence $l(xyz) > l(yz)$ if $x \neq 1$. But $l(yz) = l(y) + l(z) > l(y)$ if $z \neq 1$ (by Lemma A.0.10).

Uniqueness: Suppose $w = x_1y_1z_1 = x_2y_2z_2$ with x_i, y_i, z_i as above. Since $y_1z_1 = x_1^{-1}x_2y_2z_2$, and $y_1z_1, y_2, z_2 \in D_{W'}$, we have

$$\emptyset = I(y_1z_1) \cap W' = I(x_1^{-1}x_2y_2z_2) \cap W'$$

$$= I(x_1^{-1}x_2) \cap W' + x_1^{-1}x_2I(y_2z_2)x_2^{-1}x_1 \cap W' = I(x_1^{-1}x_2) \cap W'.$$

Hence $x_1^{-1}x_2 = 1$, i.e. $x_1 = x_2$. Thus $y_1z_1 = y_2z_2$, and hence $y_1 = y_2, z_1 = z_2$ by Proposition A.0.9.∎

The next proposition is known as the *Z-property* or *lifting property* of Bruhat order (see [Bj3], Section 2).

Proposition A.0.12 *If $x, y \in W, s \in S$ satisfy $l(sx) < l(x)$ and $l(sy) < l(y)$ then the following conditions are equivalent:*

1. $x \leq_B y$

2. $sx \leq_B y$

3. $sx \leq_B sy$.∎

The next proposition will be needed in the proof of Lemma A.0.14, and is related to Kilmoyer's Theorem ([So2], Lemma 2).

Proposition A.0.13 (Dyer) *Let $x \in D_{W'}$ and $J \subseteq S$. Then*

$$W' \cap xW_J x^{-1} = \langle S' \cap xW_J x^{-1} \rangle$$

and hence is a standard parabolic subgroup of the Coxeter system (W', S'), where S' are the canonical generators of W' (as in Proposition A.0.8).

Proof: Obviously

$$\langle S' \cap xW_J x^{-1} \rangle \subseteq W' \cap xW_J x^{-1}$$

so we only need to show that for all $y \in W' \cap xW_J x^{-1}$ we have $y \in \langle S' \cap xW_J x^{-1} \rangle$, which we do by induction on $l'(y)$. We know we can write $yx = xz$ for some $z \in W_J$. If $l'(y) = 0$ it is trivial, so assume $l'(y) > 0$ and let $s \in S' \cap I(y)$. Since $x \in D_{W'}$, $s \in I(yx) = I(xz) = I(x) + xI(z)x^{-1}$. But $s \in I(x)$ since $x \in D_{W'}$, so $s \in xI(z)x^{-1} \subseteq xW_J x^{-1}$. Thus $sy \in W' \cap xW_J x^{-1}$ and $l'(sy) < l'(x)$, so by induction $sy \in \langle S' \cap xW_J x^{-1} \rangle$. Hence $y \in \langle S' \cap xW_J x^{-1} \rangle$ as desired.∎

The final result we need is the technical lemma needed in the proof of Theorem 4.1.6.

Lemma A.0.14 (Dyer) *If $u_1, u_2, v_1, v_2 \in W$, $w \in W'$, and $J, K \subseteq S$ satisfy*

1. $u_2 u_1, v_2 v_1 \in D_{W'}$

2. $u_1 \in D_{W_J}^{-1}, u_2 \in D_{W_K}^{-1}$

3. $wv_2 v_1 = u_2 u_1 x$ *for some* $x \in W_J$

4. $wv_2 = u_2 y$ *for some* $y \in W_K$

then we have

1. $u_2 \leq_B v_2$

2. $u_2 = v_2 \Rightarrow u_1 \leq_B v_1$

This lemma may be rephrased as follows:

$$\Delta^2(W')(v_2v_1, v_2)W_{(\emptyset,\emptyset)} \subseteq \Delta^2(W')(u_2u_1, u_2)W_{(J,K)}$$

along with hypothesis 2 above implies that

$$(u_1, u_2) \leq_{\mathcal{RLB}} (v_1, v_2).$$

Proof:

Assertion 1: $v_2 = w^{-1}u_2y$, so we need to show $u_2 \leq_B w^{-1}u_2y$. *Claim:* it suffices to show $u_2 \leq Bw^{-1}u_2$. To see this, write $y = s_1 \cdots s_m$ with $s_i \in K$. Then $u_2 \leq_B w^{-1}u_2$ would imply $u_2 \leq_B w^{-1}u_2s_1$, either trivially (if $l(w^{-1}u_2s_1) > l(w^{-1}u_2)$) or by the Z-property (Proposition A.0.12) in the other case. Continuing in this way, we get that $u_2 \leq_B w^{-1}u_2$ would imply

$$u_2 \leq_B w^{-1}u_2s_1 \cdots s_m = w^{-1}u_2y.$$

Our immediate goal in proving $u_2 \leq_B w^{-1}u_2$ will be to show that

$$l'(w^{-1}u_2) = l'(w^{-1}) + l'(u_2),$$

where $l'(g) = \#I(g) \cap W'$. Since

$$I(w^{-1}u_2) = I(w^{-1}) + w^{-1}I(u_2)w,$$

this means that our goal is to show

$$(I(w^{-1}) \cap W') \cap (w^{-1}I(u_2)w \cap W') = \emptyset$$

$$\text{or } wI(w^{-1})w^{-1} \cap wW'w^{-1} \cap I(u_2) = \emptyset$$

$$\text{or } I(w) \cap W' \cap I(u_2) = \emptyset.$$

So let $t \in I(w) \cap W'$. Then $w^{-1}tw \notin I(v_2v_1)$ (since $v_2v_1 \in D_{W'}$), so

$$t \in I(w) + wI(v_2v_1)w^{-1} = I(wv_2v_1) = I(u_2u_1x) = I(u_2u_1) + u_2u_1I(x)u_1^{-1}u_2^{-1}$$

which implies $t \in u_2u_1I(x)u_1^{-1}u_2^{-1}$ (since $u_2u_1 \in D_{W'}$). Thus if we let $t = u_1^{-1}u_2^{-1}tu_2u_1$, then $t \in I(x) \subseteq W_J$. Since $u_1 \in D_{W_J}^{-1}$, we know that $l(u_1t'u_1^{-1} \cdot u_1) = l(u_1t') > l(u_1)$ and hence that $u_1t'u_1^{-1} \notin I(u_1)$. But $u_1t'u_1^{-1} = u_2^{-1}tu_2$, so $t \notin u_2I(u_1)u_2^{-1}$. Since $u_2u_1 \in D_{W'}$ implies that

$$t \notin I(u_2u_1) = I(u_2) + u_2I(u_1)u_2^{-1},$$

we must have $t \notin I(u_2)$. Therefore $I(w) \cap W' \cap I(u_2) = \emptyset$ as we wanted.

Now write $u_2 = zz'$ with $z \in W', z' \in D_{W'}$ (by Proposition A.0.9). Note that $l'(u_2) = l'(z)$ and $l'(w^{-1}u_2) = l'(w^{-1}z)$, and thus we have

$$l'(w^{-1}z) = l'(w^{-1}u_2) = l'(w^{-1}) + l'(u_2) = l'(w^{-1}) + l'(z).$$

Hence $z \leq_B w^{-1}z$ in W'. But then multiplying on the right by z' is an isomorphism of $\leq_B$ (Proposition A.0.9), so we get

$$u_2 = zz' \leq_B w^{-1}zz' = w^{-1}u_2$$

as desired.

Assertion 2: We are now assuming $u_2 = v_2$. Since

$$u_2u_1x = wv_2v_1 = u_2yv_1,$$

we have $u_1x = yv_1$ or $v_1 = y^{-1}u_1x$. Write $u_2 = v_2 = zv_3$ with $z \in W', v_3 \in D_{W'}$ (using Proposition A.0.9) and let

$$W'' = v_3^{-1}W'v_3 \cap W_K$$

(a reflection subgroup by Proposition A.0.13) . Our goal will be to show that $y^{-1} \in W''$ and $u_1 \in D_{W''}$. This would imply that u_1 is the least element of $W''u_1W_J$ (since we already have $u_1 \in D_{W_J}^{-1}$) and thus $u_1 \leq_B v_1$ by Proposition A.0.11 (since $u_1 = y^{-1}u_1x \in W''u_1W_J$).

To show that $y \in W''$, note that

$$wzv_3 = wv_2 = u_2y = zv_3y$$

and hence

$$y = v_3^{-1}z^{-1}wzv_3 \in W_K \cap v_3^{-1}W'v_3 = W''.$$

It only remains then to show that $u_1 \in D_{W''}$. We have

$$\emptyset = I(u_2u_1) \cap W' = I(zv_3u_1) \cap W' = I(z) \cap W' + zI(v_3u_1)z^{-1} \cap W'$$

and hence

$$
\begin{aligned}
I(z^{-1}) \cap W' &= z^{-1}I(z)z \cap W' \\
&= z^{-1}(I(z) \cap W')z \\
&= I(v_3u_1) \cap W' \\
&= I(v_3) \cap W' + v_3I(u_1)v_3^{-1} \cap W' \\
&= v_3I(u_1)v_3^{-1} \cap W'
\end{aligned}
$$

Thus

$$I(u_1) \cap v_3^{-1}W'v_3 = v_3^{-1}I(z^{-1})v_3 \cap v_3^{-1}W'v_3$$

and hence we have

$$I(u_1) \cap W'' = v_3^{-1}I(z^{-1})v_3 \cap W''. \tag{A.1}$$

On the other hand,

$$
\begin{aligned}
I(u_2^{-1}) \cap v_3^{-1} W' v_3 &= u_2^{-1} I(u_2) u_2 \cap u_2^{-1} W' u_2 \\
&= u_2^{-1} I(u_2 \cap W') u_2 \\
&= v_3^{-1} z^{-1} (I(z) \cap W' + z I(v_3) z^{-1} \cap W') z v_3 \\
&= v_3^{-1} I(z^{-1}) v_3 \cap v_3^{-1} W' v_3
\end{aligned}
$$

But $u_2 \in D_{W_K}^{-1}$, so $I(u_2^{-1}) \cap W_K = \emptyset$. Hence

$$
v_3^{-1} I(z^{-1}) v_3 \cap v_3^{-1} W' v_3 \cap W_K = \emptyset.
$$

This last fact, combined with equation A.1, says that $u_1 \in D_{W''}$ as we wanted.■

Bibliography

[Bj1] A. Björner, "Posets, regular CW-complexes and Bruhat order", Europ. J. Comb. **5**(1984),7-16.

[Bj2] A. Björner, "Shellable and Cohen-Macaulay partially ordered sets", Trans. Amer. Math. Soc. **260**(1980),159-183.

[Bj3] A. Björner, "Some algebraic and combinatorial properties of Coxeter complexes and Tits buildings", Adv. Math. **52**(1984), 173-212.

[Bo] N. Bourbaki, *Groupes et algèbres de Lie, Chapitres 4,5, et 6*, Éléments de mathématiques, Fasc. XXXIV, Hermann, Paris, 1968.

[Bre] K. S. Brown, *Buildings*, Graduate Texts in Mathematics, Springer-Verlag, New York, 1989.

[Ca] R. W. Carter, "Conjugacy classes in the Weyl group", Compositio Math. **25**, Fasc. 1(1972), 1-59.

[DeWa] J. Désarménien and M. Wachs, "Descentes des dérangements et mots circulaires", Actes 19e Séminaire Lotharingien de Combinatoire, Publ. IRMA Strasbourg, 1988, 13-21.

[Dy] M. Dyer, "Reflection subgroups of Coxeter systems", preprint.

[Ga] A. Garsia, "Combinatorial methods in the theory of Cohen-Macaulay rings", Adv. Math. **38**(1980), 229-266.

[GG] A. Garsia and I.Gessel, "Permutation statistics and partitions", Adv. Math. **38**(1979), 288-305.

[Ge1] I. Gessel, "Counting permutations by descents, greater index, and cycle structure", unpublished manuscript.

[Ge2] I. Gessel, "Multipartite P-partitions and inner products of skew Schur functions", Contemp. Math. **34**(1984) 289-301.

[GS] A. Garsia and D. Stanton, "Group actions on Stanley-Reisner rings and invariants of permutation groups", Adv. Math. **51**(1984), 107-201.

[HW] G. H. Hardy and E. M. Wright, *An introduction to the theory of numbers (fifth edition)*, Oxford University Press, 1979.

[HE] M. Hochster and J. A. Eagon, "Cohen-Macaulay rings, invariant theory, and the generic perfection of determinantal loci", Amer. J. Math. **93**(1971), 1020-1058.

[KM] G. Kreweras and P. Moszkowski, "Signatures des permutations et mots extraits", Discrete Math. **68**(1988), 71-76.

[Mo] P. Moszkowski, "Généralization d'une formule de Solomon relative à l'anneau de groupe d'une groupe de Coxeter", C. R. Acad. Sci. Paris, t. 309, Sér. I(1989), 539-541.

[Se] J.-P. Serre, *Linear representations of finite groups*, Graduate Texts in Mathematics 42, Springer-Verlag, New York, Heidelberg, Berlin, 1977.

[Sm] L. Smith, "Transfer and ramified coverings", Math. Proc. Camb. Phil. Soc. **93**(1983), 485-493.

[So1] L. Solomon, "The orders of the finite Chevalley groups", J. Algebra **3**(1966), 376-393.

[So2] L. Solomon, "A Mackey formula in the group ring of a Coxeter group", J. Algebra **41**(1976), 225-264.

[So3] L. Solomon, "Partition identities and invariant theory", J. Comb. Theory Ser. A **23**(1977), 148-175.

[St1] R. Stanley, *Combinatorics and commutative algebra*, Birkhäuser, Boston, 1983.

[St2] R. Stanley, "f-vectors and h-vectors of simplicial posets", preprint.

[St3] R. Stanley, "Ordered structures and partitions", Mem. Amer. Math. Soc. no. 119(1972).

[St4] R. Stanley, "Some aspects of groups acting on finite posets", J. Comb. Theory Ser. A **32**(1982), 132-161.

[Ti] J. Tits, *Spherical buildings and finite BN-pairs*, Lecture notes in mathematics Vol. 386, Springer-Verlag, Berlin and New York, 1974.

[Wa2] M. Wachs, "The major index polynomial for conjugacy classes of permutations", preprint.

Author's address:

Victor Reiner

Dept. of Mathematics

Univ. of Minnesota

Minneapolis, MN 55455

MEMOIRS of the American Mathematical Society

SUBMISSION. This journal is designed particularly for long research papers (and groups of cognate papers) in pure and applied mathematics. The papers, in general, are longer than those in the TRANSACTIONS of the American Mathematical Society, with which it shares an editorial committee. Mathematical papers intended for publication in the Memoirs should be addressed to one of the editors:

Ordinary differential equations, partial differential equations and applied mathematics to ROGER D. NUSSBAUM, Department of Mathematics, Rutgers University, New Brunswick, NJ 08903

Harmonic analysis, representation theory and Lie theory to AVNER D. ASH, Department of Mathematics, The Ohio State University, 231 West 18th Avenue, Columbus, OH 43210

Abstract analysis to MASAMICHI TAKESAKI, Department of Mathematics, University of California, Los Angeles, CA 90024

Real and harmonic analysis to DAVID JERISON, Department of Mathematics, M.I.T., Rm 2–180, Cambridge, MA 02139

Algebra and algebraic geometry to JUDITH D. SALLY, Department of Mathematics, Northwestern University, Evanston, IL 60208

Geometric topology and general topology to JAMES W. CANNON, Department of Mathematics, Brigham Young University, Provo, UT 84602

Algebraic topology and differential topology to RALPH COHEN, Department of Mathematics, Stanford University, Stanford, CA 94305

Global analysis and differential geometry to JERRY L. KAZDAN, Department of Mathematics, University of Pennsylvania, E1, Philadelphia, PA 19104-6395

Probability and statistics to RICHARD DURRETT, Department of Mathematics, Cornell University, Ithaca, NY 14853-7901

Combinatorics and number theory to CARL POMERANCE, Department of Mathematics, University of Georgia, Athens, GA 30602

Logic, set theory, general topology and universal algebra to JAMES E. BAUMGARTNER, Department of Mathematics, Dartmouth College, Hanover, NH 03755

Algebraic number theory, analytic number theory and modular forms to AUDREY TERRAS, Department of Mathematics, University of California at San Diego, La Jolla, CA 92093

Complex analysis and nonlinear partial differential equations to SUN-YUNG A. CHANG, Department of Mathematics, University of California at Los Angeles, Los Angeles, CA 90024

All other communications to the editors should be addressed to the Managing Editor, DAVID J. SALTMAN, Department of Mathematics, University of Texas at Austin, Austin, TX 78713.

General instructions to authors for

PREPARING REPRODUCTION COPY FOR MEMOIRS

**For more detailed instructions send for AMS booklet, "A Guide for Authors of Memoirs."
Write to Editorial Offices, American Mathematical Society, P.O. Box 6248,
Providence, R.I. 02940-6248.**

MEMOIRS are printed by photo-offset from camera copy fully prepared by the author. This means that the finished book will look exactly like the copy submitted. Thus the author will want to use a good quality typewriter with a new, medium-inked black ribbon, and submit clean copy on the appropriate model paper.

Model Paper, provided at no cost by the AMS, is paper marked with blue lines that confine the copy to the appropriate size.

Special Characters may be filled in carefully freehand, using dense black ink, or **INSTANT** ("rub-on") **LETTERING** may be used. These may be available at a local art supply store.

Diagrams may be drawn in black ink either directly on the model sheet, or on a separate sheet and pasted with rubber cement into spaces left for them in the text. Ballpoint pen is not acceptable.

Page Headings (Running Heads) should be centered, in CAPITAL LETTERS (preferably), at the top of the page — just above the blue line and touching it.

LEFT-hand, EVEN-numbered pages should be headed with the AUTHOR'S NAME;

RIGHT-hand, ODD-numbered pages should be headed with the TITLE of the paper (in shortened form if necessary).

Exceptions: PAGE 1 and any other page that carries a display title require NO RUNNING HEADS.

Page Numbers should be at the top of the page, on the same line with the running heads.

LEFT-hand, EVEN numbers — flush with left margin;

RIGHT-hand, ODD numbers — flush with right margin.

Exceptions: PAGE 1 and any other page that carries a display title should have page number, centered below the text, on blue line provided.

FRONT MATTER PAGES should be numbered with Roman numerals (lower case), positioned below text in same manner as described above.

MEMOIRS FORMAT

**It is suggested that the material be arranged in pages as indicated below.
Note: Starred items (*) are requirements of publication.**

Front Matter (first pages in book, preceding main body of text).

Page i — *Title, *Author's name.

Page iii — Table of contents.

Page iv — *Abstract (at least 1 sentence and at most 300 words).

Key words and phrases, if desired. (A list which covers the content of the paper adequately enough to be useful for an information retrieval system.)

*1991 Mathematics Subject Classification. This classification represents the primary and secondary subjects of the paper, and the scheme can be found in Annual Subject Indexes of MATHEMATICAL REVIEWS beginnning in 1990.

Page 1 — Preface, introduction, or any other matter not belonging in body of text.

Footnotes: *Received by the editor date.
Support information — grants, credits, etc.

First Page Following Introduction – Chapter Title (dropped 1 inch from top line, and centered). Beginning of Text.

Last Page (at bottom) – Author's affiliation.